1993

CURRICULUM AND EVALUATION

S T A N D A R D

FOR SCHOOL MATHEMATICS

ADDENDA SERIES, GRADES 9–12

A CORE CURRICULUM
MAKING MATHEMATICS COUNT FOR EVERYONE

Steven P. Meiring

Rheta N. Rubenstein

James E. Schultz

Jan de Lange

Donald L. Chambers

Consultants

Harold L. Schoen

Daniel J. Teague

Christian R. Hirsch, Series Editor

NATIONAL COUNCIL OF
TEACHERS OF MATHEMATICS

Copyright © 1992 by
THE NATIONAL COUNCIL OF TEACHERS OF MATHEMATICS, INC.
1906 Association Drive, Reston, Virginia 22091
All rights reserved

Second printing 1992

Library of Congress Cataloging-in-Publication Data:

Meiring, Steven P.
  A core curriculum : making mathematics count for everyone /
Steven P. Meiring ; with Donald L. Chambers ... [et al.].
    p.   cm. — (Curriculum and evaluation standards for school
mathematics addenda series. Grades 9–12)
  Includes bibliographical references.
  ISBN 0-87353-309-7 (set). — ISBN 0-87353-328-3 (vol.) : $17.00
  1. Mathematics—Study and teaching (Secondary) 2. Education—
Curricula. I. National Council of Teachers of Mathematics.
II. Series.
QA11.M482   1992
                                                        92-6996
                                                        CIP

Printed in the United States of America

510.712
M514

TABLE OF CONTENTS

146, 516

FOREWORD

As the *Curriculum and Evaluation Standards for School Mathematics* (NCTM 1989) was being developed, it became apparent that supporting publications would be needed to aid in interpreting and implementing the curriculum and evaluation standards and the underlying instructional themes. A Task Force on the Addenda to the *Curriculum and Evaluation Standards for School Mathematics,* chaired by Thomas Rowan and composed of Joan Duea, Christian Hirsch, Marie Jernigan, and Richard Lodholz, was appointed by Shirley Frye, then NCTM president, in the spring of 1988. The Task Force's recommendations on the scope and nature of the supporting publications were submitted in the fall of 1988 to the Educational Materials Committee, which subsequently framed the Addenda Project for NCTM Board approval.

Central to the Addenda Project was the formation of three writing teams to prepare a series of publications targeted at mathematics education in grades K–6, 5–8, and 9–12. The writing teams consisted of classroom teachers, mathematics supervisors, and university mathematics educators. The purpose of the series was to clarify the recommendations of selected standards and to illustrate how the standards could realistically be implemented in K–12 classrooms in North America.

The themes of problem solving, reasoning, communication, and connections have been woven throughout each volume in the series. The use of technological tools and the view of assessment as a means of guiding instruction are integral to the publications. The materials have been field tested by teachers to ensure that they reflect the realities of today's classrooms and to make them "teacher friendly."

We envision the Addenda Series being used as a resource by individuals as they begin to implement the recommendations of the *Curriculum and Evaluation Standards.* In addition, volumes in a particular series would be appropriate for in-service programs and for preservice courses in teacher education programs.

On behalf of the National Council of Teachers of Mathematics, I would like to express sincerest appreciation to the authors, consultants, and editor, who gave willingly of their time, effort, and expertise in developing these exemplary materials. Special thanks are due to Timothy V. Craine, who provided the authors with insights on curricular options and useful ideas for engaging lessons. Gratitude is also expressed to the following teachers who reviewed drafts of the material as this volume progressed: James Huber, John Janty, Catherine Sanor, and Edna Vasquez. Finally, the continuing technical assistance of Cynthia Rosso and the able production staff in Reston is gratefully acknowledged.

Bonnie H. Litwiller
Addenda Project Coordinator

PREFACE

The *Curriculum and Evaluation Standards for School Mathematics*, released in March 1989 by the National Council of Teachers of Mathematics, has focused national attention on a new set of goals and expectations for school mathematics. This visionary document provides a broad framework for what the mathematics curriculum in grades K–12 should include in terms of content priority and emphasis. It suggests not only what students should learn but also how that learning should occur and be evaluated.

Although the *Curriculum and Evaluation Standards* specifies the key components of a quality contemporary school mathematics program, it encourages local initiatives in realizing the vision set forth. In so doing, it offers school districts, mathematics departments, and classroom teachers new opportunities and challenges. This two-edged sword of both opportunity and challenge is perhaps nowhere more sharply defined than in the call for a core curriculum at the high school level. In contrast to current practice of prematurely tracking students into either college preparatory sequences or "general mathematics" sequences on the basis of narrow perceptions of performance or curricular goals, the *Standards* recommends at least three years of mathematical study for every student "revolv[ing] around a core curriculum differentiated by the depth and breadth of the treatment of topics and by the nature of applications." Thus, the core topics of high school mathematics are to be fundamentally the same for *all* students.

The purpose of this volume, and others in the Addenda Series, is to provide instructional ideas and materials that will support implementation of a core curriculum in local settings. It addresses in a very practical way the content, pedagogy, and pupil assessment dimensions of reshaping school mathematics to this end. It also focuses on issues that must be addressed in a transition to a core curriculum.

Reshaping Content

The curriculum standards for grades 9–12 identify a common core of mathematical topics that *all* students should have the opportunity to learn. The need to prepare students for the workplace, for college, and for citizenship is reflected in a broad mathematical sciences curriculum. The traditional strands of algebra, functions, geometry, and trigonometry are balanced with topics from data analysis and statistics, probability, and discrete mathematics. The broadening of the content of the curriculum is accompanied by a broadening of its focus. Narrow curricular expectations of memorizing isolated facts and procedures and becoming proficient with by-hand calculations and manipulations give way to developing mathematics as a connected whole with an emphasis on conceptual understanding, multiple representations and their linkages, mathematical modeling, and problem solving.

As envisioned in the *Curriculum and Evaluation Standards*, investigation of patterns, data analysis, and modeling serve to connect mathematics to the world in which students live. Coordinate representations connect data analysis with algebra, algebra with geometry, and geometry with trigonometry. Computer and calculator graphics not only bring these connections to the forefront of the curriculum but also allow students to investigate connections in a dynamic way. Moreover, the visualization approach offered by these technologies promises to afford more students greater access to mathematics.

Computer-based storage, manipulation, and retrieval of data, together with the wide use of computer spreadsheets, have increased the importance of matrix representations and methods. The topic of matrices occurs repeatedly throughout the high school standards and, as such, serves as another link between the algebra and geometry strands and in turn between these strands and the discrete mathematics strand. Chapter 2 in this volume provides a fresh perspective on matrices based on a truly innovative program for the majority of high school students in the Netherlands.

A Core Curriculum: Making Mathematics Count for Everyone offers several possible curriculum models for organizing the mathematics content recommended in the *Curriculum and Evaluation Standards.* A Crossover Model is elaborated in chapter 3, and a complete syllabus for this model appears in Appendix I. The Crossover Model represents what might be a first step away from current multitrack programs that offer advanced mathematics for a few and minimal mathematics for the majority. This transition model features two parallel sequences of courses with common treatment of core topics at each of the first three years. Chapter 3 is replete with examples of abbreviated lessons that suggest how such a model could be implemented in today's classrooms.

More robust models of a core curriculum are presented in chapter 4. Both the Enrichment Model and the Differentiated Model consist of a single sequence of courses. Appendix II provides a possible four-year syllabus for the Differentiated Model—the model that most closely approximates the curriculum organization envisioned in the *Curriculum and Evaluation Standards.* Each year of the syllabus features connected content drawn from four strands: algebra/functions, geometry/trigonometry, statistics/probability, and discrete mathematics/calculus underpinnings. Prototype lessons illustrating how such a model could be operationalized appear in chapter 4.

The curriculum models and related syllabi and sample lessons assume the appropriate use of graphing utilities, geometry exploration software, data analysis software, and spreadsheets. The annotated bibliography at the end of this volume provides additional information on computer software tools as well as on sources of ideas and materials for beginning the transition to a core curriculum.

Reshaping Pedagogy

The *Curriculum and Evaluation Standards* paints mathematics as an activity and a process, not simply as a body of content to be mastered. Throughout, there is an emphasis on doing mathematics, recognizing connections, and valuing the enterprise. Hence, standards are presented for Mathematics as Problem Solving, Mathematics as Communication, Mathematics as Reasoning, and Mathematical Connections. The intent of these four standards is to frame a curriculum that ensures the development of broad mathematical power in addition to technical competence; that cultivates students' abilities to explore, conjecture, reason logically, formulate and solve problems, and communicate mathematically; and that fosters the development of self-confidence.

Realization of these process and affective goals will require, in many cases, new teaching-learning environments. The traditional view of the teacher as authority figure and dispenser of information must give way to that of the teacher as catalyst and facilitator of learning. To this end, the standards for grades 9–12 call for increased attention to—

♦ actively involving students in constructing and applying mathematical ideas;

- using problem solving as a means as well as a goal of instruction;
- promoting student interaction through the use of effective questioning techniques;
- using a variety of instructional formats—small cooperative groups, individual explorations, whole-class instruction, and projects;
- using calculators and computers as tools for learning and doing mathematics.

A Core Curriculum: Making Mathematics Count for Everyone reflects the new methodologies supporting new curricular goals. The sample instructional activities in chapters 2, 3, and 4 provide tasks and problem situations that require students to experiment, collect data, search for patterns, make conjectures, and verify discoveries. These activities are ideally suited to cooperative group work. Differentiation in learning outcomes occurs by blending core lessons for all students with extended activities that students can complete to different depths and levels of abstraction and formalism. As should be the case with all student investigations, provisions are made for students to share their experiences, clarify their thinking, generalize their discoveries, and construct convincing arguments. Teacher-moderated discussions of student learnings would provide opportunities for all students both to review their understandings of a topic at a range of levels of abstraction and to have exposure to a range of approaches to problem-solving applications.

Teaching Matters, a special margin feature of this book, furnishes helpful suggestions on how instruction can be adapted to the thinking of students and to the resources available. Ideas for introducing topics and effectively using technological tools are also featured in these margin notes.

Reshaping Assessment

Complete pictures of classrooms in which the *Curriculum and Evaluation Standards* is being implemented not only show changes in mathematical content and instructional practice but also reflect changes in the purpose and methods of student assessment. Classrooms where students are expected to be engaged in mathematical thinking and in constructing and reorganizing their own knowledge require adaptive teaching informed by observing and listening to students at work. Thus informal, performance-based assessment methods are essential to the new vision of school mathematics.

Analysis of students' written work remains important. However, single-answer paper-and-pencil tests are often inadequate to assess the development of students' abilities to analyze and solve problems, make connections, reason mathematically, and communicate mathematically. Potentially richer sources of information include student-produced analyses of problem situations, solutions to problems, reports of investigations, and journal entries. Moreover, if calculator and computer technologies are now to be accepted as part of the environment in which students learn and do mathematics, these tools should also be available to students in most assessment situations.

A Core Curriculum: Making Mathematics Count for Everyone reflects the multidimensional aspects of student assessment and the fact that it is integral to instruction. *Assessment Matters,* another special margin feature, provides suggestions for assessment techniques focusing on students' problem solving, reasoning, and disposition toward mathematics as well as on their understanding of content. Chapter 2 provides an overview of some promising new assessment practices in the Netherlands, which are worthy of further study.

Conclusion

As indicated in the *Curriculum and Evaluation Standards,* the recommendation for a three-year core curriculum for all students is the most fundamental change proposed for grades 9–12. To provide a more extensive and a qualitatively different mathematical education to more students than ever before poses a significant challenge to schools. To fully realize the goals of a common core curriculum for all students will require sustained collegial efforts on the part of both teachers and school administrators.

The transition to a core curriculum will require changes in many areas—beliefs and attitudes of students, colleagues, and supporting publics, methods of teaching, assessment practices, and instructional materials and resources, to name a few. To wait for any of these changes to be completely in place before beginning is both unrealistic and undesirable. There will be no single template for implementation. As underscored in chapter 5, change is a process, not an event; it occurs gradually and requires significant amounts of time for reflection and accommodation on the part of those involved.

The process of change that we as mathematics educators go through incrementally—reexamining our teaching, challenging our beliefs, revising student goals, redesigning curricula, exploring alternative methods of evaluation, integrating technology into instruction—may be the most significant impetus for reshaping attitudes and fostering support at the local level for a core curriculum. Collectively, our efforts will achieve a grassroots redefinition of the contribution that mathematics makes to the education of *everyone.*

Christian R. Hirsch, Editor
Grades 9–12 Addenda Series

CHAPTER 1
MATHEMATICS IN A CHANGING WORLD

In time of change, the world belongs to those who can grasp the nature of that change and fashion their life and culture to make the most of it.

—James D. Finn

The world is in the midst of very rapid change—changes in the political structures of nations; shifts from military power to economic leverage in achieving national ends; and globalization of issues affecting the environment, the spread of disease, and the response to social causes. Traditional views are being reassessed and altered. Resources, both material and human, are being reallocated and thought of in new ways. Technology, scientific advances, and population-related phenomena are raising new challenges that require significantly different thinking.

It is against this background of change that reform in mathematics education is occurring in this country and around the world. That this reform is revolutionary in character and scope is the first challenge to our thinking. Unlike evolutionary change, restructuring needs to begin with a fundamental reexamination of the purposes of mathematics education and the role that mathematical knowledge plays in our society. In this chapter we try to set into context three of the many and diverse factors that are driving the need for reform in the ways that we think about and approach mathematics instruction: (1) the changing nature of the discipline of mathematics; (2) changes in society's emphasis on mathematics; and (3) changes in our understanding of teaching and learning mathematics.

CHANGES IN THE NATURE OF MATHEMATICS

For decades the high school mathematics curriculum has been designed to provide a background suitable for studying calculus. This classical preparation in algebra, geometry, and precalculus mathematics has been based on the notion that calculus is a powerful tool in the study of further mathematics and other applied fields. The functions emphasized in high school courses are well-behaved continuous functions, easily graphed and amenable to differentiation and integration in later study. Much of the twentieth century can be characterized by the successful application of continuous mathematics to various fields of human endeavor, particularly in engineering and in the efforts of physical scientists to understand the forces of nature in terms of fundamental building blocks and interactions.

Expanding the Newtonian Curriculum

A need to enlarge the scope of this "Newtonian" mathematics curriculum began to emerge as mathematics increasingly became a tool in the social sciences. Describing and making sense of large amounts of data, drawing inferences from and making predictions on the basis of samples of a population, and applying the ideas of probability and combinatorics to natural phenomena and human endeavors require methods from statistics and discrete mathematics.

As the application of mathematics to real-world problems has grown increasingly complex, the tools associated with the traditional curriculum have become limiting. Situations that can be represented or approximated with linear relationships are easily dealt with. But complex problems, such as making decisions about oil production or designing transportation routes, require linear systems with tens of thousands of equations and variables. Other applications can only be described with nonlinear systems, which generally cannot be solved. Or they require higher-degree functions whose zeroes cannot be found by purely analytical methods. Finding solutions to problems of this nature requires numeric methods of approximation, Monte Carlo methods of modeling, and computer simulations.

Whereas investigators once relied primarily on their creativity and the sophistication of known mathematical methods to guide them in the solution of problems, technology now provides capabilities that alter both the form and the means of solution. The power of calculators and computers to approximate solutions to complex problems through repetitive operations not only extends the analytical tools of the investigator but also changes the methods and areas of research.

For example, the applications of mathematical models of behavior on a real-world scale have found mixed success. Sometimes the behavior proves to be periodic and matches the models closely. But at other times, the behavior becomes aperiodic and unpredictable. Small changes in conditions produce enormous changes in outcome in the next occurrence—a tornado occurs when weather conditions vary only slightly from normal. *Difference equations,* $x_{next} = F(x)$, are one example of iterative mathematical techniques suitable for modeling behavior that jumps from one state to another, such as the way an animal population changes from year to year.

Technology also makes the separation of areas of study in mathematics less clear. As the importance of shape, pattern, and self-similarity grows in the applications of mathematics, geometry plays an ever-growing role as a mathematical means to permit scientists and engineers to visualize the results of their models. Whereas the Cartesian coordinate system once gave us the means to visualize the relationships conveyed by equations like the conic sections, the use of computer-generated graphs gives the mathematician the means to visualize what occurs in iterative processes. Geometry enables the investigator to explore the formation of shapes and to describe them in terms of dimension and scaling.

A Mathematics Curriculum for the Future

The implications for school mathematics of the changes occurring in mathematics and its applications are likely to be far-reaching. The study of topics from probability and statistics has assumed increasing importance to the fields of behavioral and social sciences. Discrete mathematics, the mathematics of dynamical systems (chaos), and mathematical computing must supplement the traditional precalculus curriculum to provide an accurate background and understanding of the way that mathematics is applied to real-world problems. The growing connections among mathematical topics and with other fields raise serious questions about the way that we package the high school curriculum into discrete subjects—making us one of the few industrialized countries of the world that do not organize and teach mathematics as an integrated discipline.

CHANGES IN SOCIETY'S EMPHASIS ON MATHEMATICS

American society has long held the view that sustained instruction in mathematics and science is suitable only for the most academically talented. As long as some small segment of the population became highly trained in these fields, we have believed that our needs as a society would be met in maintaining our positions of leadership in industry, technology, scientific enterprise, and world affairs. In an industrialized society, lack of mathematical expertise was not much of an impediment to success in the workplace. Tasks were largely arithmetical; if more complex, they required the attention of a specialist in the application of mathematics—an engineer, an accountant, a statistician, a quality control analyst. Production was standardized through assembly lines, and workers were not expected to participate in decisions. Work requiring the application of mathematical ideas occurred mostly in the trades, and such applications were learned on the job.

Understanding social and political issues was similarly uncomplicated by mathematical sophistication. Applying mathematics for personal needs was generally limited to arithmetic and measurement. The mathematics learned in the elementary grades was essentially the level of competence required for later life. High school coursework served to teach students how to reason or to prepare them for post–high school studies. It bore little resemblance to how mathematics was applied in the real world, except by skilled professionals.

Needs in a Postindustrial Society

Lack of competence in mathematics beyond arithmetic now limits an individual's opportunity for success in life and a nation's economic strength and leadership potential. Virtually every area of life requires a higher competence with mathematics for full participation in society. The ability to understand significant mathematics and mathematical procedures is necessary for making informed judgments on issues, acting as a wise consumer, and coming to personal and business decisions. Lack of success in high school level mathematics and beyond now eliminates graduates from all but the most menial dead-end jobs. Failure to produce a work force trained to compete against the quality of workers in Pacific Basin countries and western European nations threatens to erode our national capacity to compete economically, adversely affecting our standard of living.

Signposts of Increased Need for Mathematical Literacy

- After demographic factors, the strongest predictor of earnings nine years after graduation from high school is the number of mathematics courses taken.

- Labor experts estimate that few jobs will remain the same for longer than five years. High school graduates need the knowledge and skills to pursue a series of careers, with the expectation that they will face a lifetime of learning.

- Natural career ladders are vanishing. New technology has increasingly separated "back office" functions (clerical, service) from "front office" functions (technical, sales, professional). It will become increasingly difficult to work up the ranks through informal on-the-job training.

- Stable employment in manufacturing, communications, transportation, utilities, and forestry—high-paying sectors once wide open to young people under age 20—has fallen from 57 percent in 1968 to 36 percent in 1986.

- With fewer new, young workers entering the work force, employers will be more likely to offer jobs and training to those they have traditionally ignored. But minority workers are less likely to have had the requisite schooling and on-the-job training. Language, attitude, and cultural factors may also prevent access to these jobs. Without significant changes, African-Americans and Hispanics will have a smaller fraction of jobs in the year 2000 than today, whereas their percentage of those seeking work will have risen.

◆ ◆ ◆ ◆ ◆ ◆ ◆ ◆

*Everything used to be me-
chanical, and if it didn't work
you got a bigger hammer. But
those days are over.*
    *—Dennis Walsh, director of
    training for Swift Textiles*

Even in the most labor-intensive industries, the industrial model of the
world of work has dramatically altered. Assembly-line methods and
physically demanding tasks are being replaced by computer technology
and automation. Workers who passively perform the same daily tasks
under close supervision are being replaced by operators who make
many of their own decisions, monitor their own quality output, and work
in small teams of employees who rotate jobs. International competition
places a premium on customized production, innovation, rapid infusion
of technology, and product quality. Decision making and responsibility
are being placed back on the individual with a resulting de-emphasis on
middle management. Fast decisions require workers who can interpret
data, read technical manuals, communicate with other team members,
and undergo continuous retraining.

The success of today's industries depends on the availability of a vast
labor pool of skilled and adaptable workers who possess process skills
in reasoning and interpersonal relations, as well as in mathematics,
communications, aptitudes, and the work ethic. (See fig. 1.1.) The educational system for preparing these future workers must have vastly
different goals and methods than that for the industrial age. Other
countries have recognized the role of education in developing an infra-
structure that supports the technology, the work force, and the scientific
needs of their economies. The rebirth of Europe and Japan after World
War II attests that the foundation of national wealth is really people—the
human capital represented by their knowledge, skills, organizations, and
motivations. The income-generating assets of a nation are the knowledge
and skills of its workers.

Individual Opportunity Linked to Education

An industrial age supported multiple ways of achieving a high standard
of living. Those who pursued jobs in manufacturing after high school
found a career ladder for advancing to middle management or trade
positions with minimal additional training. Pay was commensurate with
that of professional positions requiring advanced education. A post-
industrial economy has fewer high-paying production jobs, requires fewer
tradespeople, has eliminated much of middle management, and has replaced production jobs with lower-paying service jobs. The median years
of education for the new jobs that will be created between now and the
year 2000 will be 13.5 years. For the first time in our history, the
majority of new jobs require postsecondary education.

Families with two wage earners, workers employed in more than one
job, and young people who delay marriage and family are indications
of how individual society members are trying to adjust their life-styles to
maintain a desirable standard of living. But there is a limit to maintaining
prosperity in these ways. Between 1979 and 1985, 1.7 million jobs
were lost in manufacturing. Millions of new jobs were created in the retail
and service sectors, but wages typically were only half those in manufacturing. The fastest-growing jobs are in professional, technical, and sales
fields requiring advanced education and high skill levels. Advancement in
nearly every field in the future will require further education. Unless our
educational system raises the standards for *all* students and prepares
each for lifelong learning, we stand in danger of becoming a nation
divided by education.

*What the future holds in
store for individual human
beings, the nation, and the
world depends largely on the
wisdom with which humans
use science and technology.
But that, in turn, depends on
the character, distribution,
and effectiveness of the
education that people receive.*
    —Project 2061

Maintaining Democratic Values

There is more at issue than our economic self-interest. Education is
closely linked to preserving our democratic ideals. A common core of
knowledge, skills, and values is necessary to maintain our beliefs in the
underpinnings of our society. The growth of an underclass of the long-

A Need for Greater Skills

Occupations of the future will require higher skill levels, based on the Labor Department's breakdown of the skills needed by workers to perform a wide range of jobs. In 1984, for instance, 6% of jobs required workers with the two highest skill levels; for jobs to be created between 1984 and 2000, that figure will rise to 13%.

DEFINING SKILL LEVELS

SKILL LEVEL	LANGUAGE SKILL LEVEL	MATH SKILL LEVEL
6	Reads literature, book and play reviews, scientific and technical journals, financial reports and legal documents. Writes novels, plays, editorials, speeches, critiques.	Advanced calculus, modern algebra and statistics.
5	Same as level 6, but less advanced.	Knows calculus and statistics; econometrics.
4	Reads novels, poems, newspapers, manuals, thesauri and encyclopedias. Prepares business letters, summaries and reports. Participates in panel discussions and debates. Speaks extemporaneously on a variety of subjects.	Is able to deal with fairly complex algebra and geometry, including linear and quadratic equations, logarithmic functions and deductive axiomatic geometry.
3	Reads a variety of novels, magazines and encyclopedias, as well as safety rules and equipment instructions. Writes reports and essays with proper format and punctuation. Speaks well before an audience.	Understands basic geometry and algebra. Calculates discount, interest, profit and loss, markup and commissions.
2	Recognizes meaning of 5,000-6,000 words. Reads at a rate of 190-215 words per minute. Reads adventure stories and comic books, as well as instructions for assembling model cars. Writes compound and complex sentences, with proper end punction and using adjectives and adverbs.	Adds, subtracts, multiplies and divides all units of measure. Computes ratio, rate and percent. Draws and interprets bar graphs.
1	Recognizes meaning of 2,500 (two- or three-syllable) words. Reads at rate of 95-120 words per minute. Writes and speaks simple sentences.	Adds and subtracts two-digit numbers. Does simple calculations with money and with basic units of volume, length and weight.

EXISTING JOBS   NEW JOBS   👤 = 1%

SKILL LEVEL NEEDED

OCCUPATION	LANGUAGE	MATH
Biochemist	6	6
Computer-applications engineer	6	6
Mathematician	6	6
Cardiologist	6	5
Social psychologist	6	5
Lawyer	6	4
Tax attorney	6	4
Newspaper editor	6	3
Accountant	5	5
Personnel manager	5	5
Corporate president	5	5
Weather forecaster	5	5
Secondary school teacher	5	4
Disk jockey	5	3
Elementary school teacher	5	3
Financial analyst	4	5
Corporate vice president	4	5
Computer-sales representative	4	4
Management trainee	4	4
Insurance-sales agent	3	4
Retail-store manager	3	4
Cement mason	3	3
Manager of dairy farm	3	3
Poultry farmer	3	3
Tile setter	3	3
Travel agent	3	3
Directory-assistance operator	3	2
Janitor	3	2
Short-order cook	3	2
Assembly-line worker (appliances)	2	2
Toll collector	2	2
Laundry worker	1	1

Source: The Labor Department

EDUCATION

▨ New Jobs (created 1985-2000)
█ Existing Jobs (1985)

4 years of college or more
30%
22

1-3 years of college
22
20

4 years of high school
35
40

3 years of high school or less
14
18

Median years of school
13.5 years
12.8

Source: The Hudson Institute

KARL HARTIG

Fig. 1.1

term unemployed, dominated by particular subcultures, is intolerable. Aside from the attendant welfare dependency, crime, and social unrest, the notion of large segments of our society limited in their opportunities to participate fully in the mainstream of American life runs contrary to the principles that our nation represents.

Equally at risk is our capacity to govern ourselves wisely and to act responsibly as individuals. The problems of a modern technological society are complex, requiring an electorate who can sift through arguments, interpret quantitative information, make critical judgments, and look beyond immediate self-interest. The abilities to reason and to think and act independently are survival skills for being wise consumers, good stewards of the environment, intelligent supporters of rational policies of government, and citizens capable of appreciating cultural differences.

Mathematical literacy to function in a technological society can no longer be the goal of an elite subset of the school population. Every student must be equipped with the knowledge and skills to make sense of data, to interpret technical materials, to understand linear and nonlinear growth, to manipulate formulas and algebraic symbols, to distinguish logical arguments, to appreciate and act on uncertainty, and to apply geometric principles. Each individual must be equipped with a combination of personal skills, technological skills, and thinking skills in order to apply mathematics meaningfully. These are the prerequisites for understanding the world in which we live, for realizing the potential of technology, and for maintaining our system of government.

CHANGES IN THE TEACHING AND LEARNING OF MATHEMATICS

We have learned a great deal in the last few decades about how students learn mathematics. New approaches to instruction and assessment make it possible to increase the scope and depth of the study of mathematical topics for wider and more diverse student populations.

Active Learning

Learning in mathematics is synonymous with doing—*predicting and verifying, generalizing, finding and expressing patterns, modeling, visualizing, conjecturing, linking ideas*—not by the teacher but by the student. When students are active participants in helping to shape their own understanding of an idea, more happens than merely recognizing the resulting technique or principle. The learner is learning how to learn.

Consider, for example, a typical instructional episode involving the law of sines. A passive introduction to this relationship might first present an acute triangle (fig. 1.2). The formula is quickly derived by drawing an altitude from one vertex. To complete the development, both the right triangle and the obtuse triangle cases are considered. The lesson then moves on to introductory examples and applications of this result.

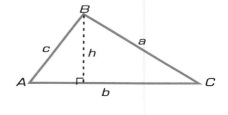

$$h = c \sin A = a \sin C$$
$$\frac{c}{\sin C} = \frac{a}{\sin A}$$

Fig. 1.2

Missing in this development is the active involvement of the learner in first exploring whether such a relationship exists and *then* setting about to establish it. Consider an alternative dialog with students that raises the following questions (fig. 1.3):

Consider triangles in which two sides are held constant while the measure of the included angle varies. As the measure of this angle increases or decreases, what happens to the measure of the side opposite? How can we determine whether a direct variation is occurring?

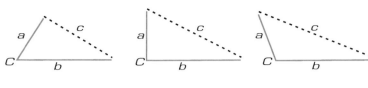

Fig. 1.3

Discussion may lead to a review of the fact that direct variation is represented by the form $y = kx$, or the constant quotient $\frac{y}{x} = k$. As students consider quotients that involve an angle of a triangle and the side opposite, several candidates for a direct variation relationship emerge. Below are three possibilities.

$$\frac{side}{angle} \qquad \frac{side}{sine\ of\ angle} \qquad \frac{side}{tangent\ of\ angle}$$

Now, triangles should be sketched and measurements taken by students to test these possibilities. For the triangle in figure 1.4, the resulting table might look like the following:

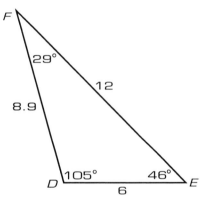

Fig. 1.4

side	angle	$\frac{side}{angle}$	$\frac{side}{sine\ of\ angle}$	$\frac{side}{tangent\ of\ angle}$
a	A	6.5	12.4	–3.2
b	B	11.1	12.4	8.6
c	C	11.9	12.4	10.8

As students generalize the apparent relationship for this particular triangle, they complete a far more important activity than merely establishing that such a law may exist (still to be proved). They learn how mathematics is discovered, how to formulate and test conjectures, and how to state generalizations. Their resulting insight into the meaning of the law of sines will be far deeper than that of solving a proportion. Students are more likely to think of related questions that lead to further investigation.

Mathematics is a tool for answering questions that are meaningful to learners. As we achieve a balance between the deductive, formal parts of mathematics and the intuitive, experiential elements of its study, we find that mathematics makes more sense and has relevance for more and more students. Simulations and modeling, in particular, make it possible to study situations in a less abstract way.

Smart is not something that you just are; smart is something that you can get.
—Jeff Howard,
Getting Smart: The Social Construction of Intelligence

Cooperative Learning

The activity of students learning in small cooperative groups models the way that mathematics is applied in today's work world. More important, it is an effective means of improving learning and extending understanding among more diverse groups. The following activity (fig. 1.5) is characteristic of a cooperative learning project, designed to actively engage students over three to four days. Students are expected to work in groups of four or five students to explore analytically and then experientially a question related to target selection in dart throwing. (Blackline masters for this activity are on pages 13–16.)

DART THROWING   SHEET 1

Figures A through E represent targets for dart throwing.

A   B   C   D   E

Suppose that you can earn points by throwing darts according to these rules:

You score 1 point, if your dart lands inside a circle.

You score 0 points, if your dart lands inside a square but outside a circle.

Darts thrown outside a square do not count (you can throw them again.)

Which target would you choose to throw at? _____

Why?

How can you support your answer mathematically?

Hint 1: Think about the combined areas of circles within a square and the area of that square.

Hint 2: What is a "nice" number to choose for the side length of the squares in order to perform the calculations easily?

Describe your plan in words.

DART THROWING (CONTINUED)   SHEET 2

Leave all results below in terms of π.

Target A

Radius of one circle: _____ units

Area of one circle: _____ square units

Ratio: _____ = _____
       (state in words)   (value)

Information for All Targets
Side length of square: _____ units
Area of square: _____ square units

Target B

Radius of one circle: _____ units

Area of one circle: _____ square units

Area of _____ circles: _____ square units

Ratio: _____ = _____
       (state in words)   (value)

Target C

Target D

Target E

What do you conclude?

Express the area ratio(s) for all targets as a decimal. What fraction or percentage of the time would you expect a dart to land inside a circle if you knew that it hit inside the square?

DART THROWING (CONTINUED)   SHEET 3

Experimental Results

Part I   Your teacher will supply target A and another target.

Choose a member of your group to be the dart thrower. Measure a distance eight feet from the target and place a piece of tape on the floor. Standing behind this tape, the dart thrower throws some number of times (your choice) at target A. Keep a tally of darts (a) inside the circles and (b) in the square but outside any circle. Darts outside the square do not count.

CAUTION: All members of the group must stay behind the thrower, and only the thrower retrieves the darts.

Repeat this activity with another target supplied by your teacher.

For each target, calculate as a decimal the ratio of the darts inside the circles to the total darts hitting the target.

Target A	Target _____
$\dfrac{\textit{darts inside circles}}{\textit{darts hitting target}}$ =	$\dfrac{\textit{darts inside circles}}{\textit{darts hitting target}}$ =
ratio:	ratio:

Are these ratios the same? Are they about the same as your "mathematical" results on sheet 2? What do you conclude?

What decision(s) did you have to make as you did the activity? Might they have influenced your ratios? Why or why not?

DART THROWING (CONTINUED)   SHEET 4

Experimental Results

Part II   You will use the same two targets for this activity that you used for part I.

Choose another member of your group to be the dart thrower. Measure a distance six feet from the target and place a piece of tape on the floor. Standing behind this tape, the dart thrower *with eyes closed* throws some number of times (your choice) at target A. Counting only darts that hit the target, keep a tally of darts (a) inside the circles and (b) outside the circles.

CAUTION: All members of the group must stay behind the thrower, and only the thrower retrieves the darts.

Repeat this activity with another target supplied by your teacher.

For each target, calculate as a decimal the ratio of the darts inside the circles to the total darts hitting the target.

Target A	Target _____
$\dfrac{\textit{darts inside circles}}{\textit{darts hitting target}}$ =	$\dfrac{\textit{darts inside circles}}{\textit{darts hitting target}}$ =
ratio:	ratio:

Are these ratios the same? Are they about the same as those computed in part I? Are they about the same as your "mathematical" results on sheet 2? What do you conclude?

Fig. 1.5

This activity is structured to appeal to students' intuition in making a real-world choice. Their intuitive approach is then extended to a mathematical analysis of the situation, followed by experiential data collecting that results in an incongruity between predicted and actual results. (Mathematical analysis indicates that the targets are equivalent for dart throwing, but data reveal that Target A is superior for maximizing a score.) Ensuing class and small-group discussions to resolve the apparent contradiction between the area model for target selection and the empirical results reveal a great deal about *modeling, assumptions, randomness,* and *the need for verification.*

Critical to the communication goals for the activity is the requirement for students to record their thinking: (1) the rationale for their "hunch"; (2) their mathematical approach to comparing targets; (3) the results of their data study; and (4) their conclusions. Throughout the project, students work cooperatively, discuss freely, and arrive at group results. Learning is noncompetitive. However, each student is expected to write a personal report, thereby assuring individual accountability, growth in written communication skills, and consolidation of learning. Individual outcomes are expected.

A project focus helps students develop the skills to apply mathematics to more realistic situations. Discretely studied skills and concepts must be integrated successfully within the same learning activity. Extended learning time over several days permits students to engage in group problem solving and to learn from one another. Cooperative learning encourages students to grapple with content and topics that might prove too formidable for individually oriented study, such as lecture-recitation.

Technology-supported Learning

Effective instruction takes advantage of both group and individual learning. Technology, in particular, permits instruction to become more diversified and individualized. Instructional software extends teachers' latitude in varying the pace and the reinforcement needed to match students' learning rates and styles. While part of a class is engaged in a broadening or extension activity associated with a core topic, tutorial software could be used to solidify understanding of that topic for individual students.

Perhaps the most exciting potential of technology, however, is its effect on increasing the amount of time that can be devoted to developing conceptual understandings and reasoning processes that lie at the heart of mathematical problem solving. Spreadsheets permit students to explore situations without excessive algebraic manipulation. Calculators and computers remove the onerous and time-consuming manipulative aspects of an investigation. Graphing utilities and instructional software like the Geometric Supposer or Geometer's Sketchpad enable students to visualize relationships readily and to test ideas quickly. Statistical packages enable real data from real experiments to be analyzed using readily produced visual displays, summary statistics, and prediction models.

The resulting de-emphasis on the need to develop intricate manipulative skills (NCTM 1989, p. 127) creates room in the curriculum for other topics made relevant and accessible by the application of technology (NCTM 1989, p. 126). Effective use of technology, however, presumes a new set of skills, including key sequencing, scaling, zoom-in, zoom-out, domain and range settings, and cell definition. When we consider how mathematics is applied in the work world at even entry-level positions, such technology-associated skills realistically must be expected of all students.

Numbers are to a mathematician what bags of coins are to an investment banker; nominally the stuff of his profession, but actually too gritty and particular to waste time on. Ideas are the real currency of mathematicians.
—James Gleick, Chaos

The Hydramatic Transmission plant in Toledo, Ohio, employs 4800 persons. There are 600 personal computers on the production floor of the plant. Each employee gets forty hours of training in statistical quality control and is expected to monitor material use and cost controls through CRTs at the job location, using graphing utilities to create bar and circle graphs as one means of displaying this information. $4.5 million each year is allocated for nonprofessional training—an average of forty hours for each employee each year.

van Hiele Model

1. *Recognition—students view figures as entities.*

2. *Analysis—students recognize characteristics of figures.*

3. *Ordering—students see interrelationships among figures and class inclusions.*

4. *Deduction—students understand the role of undefined terms, axioms, definitions, postulates, and theorems; they can construct proofs.*

5. *Rigor—students can operate in and appreciate alternative axiomatic systems.*

Developmental Learning

The nature of learning materials is rapidly evolving to reflect our discoveries of how students acquire mathematical ideas. Consideration is given to *connecting* the development of new ideas to what the student already knows. For example, current geometry materials give attention to the five van Hiele levels of understanding geometric ideas. Student assessments help teachers determine how to plan a lesson to develop the ideas from the students' base of understanding. Learning activities keyed to particular levels of understanding then offer teachers multiple options in prescribing appropriate learning tasks.

Similarly, promising materials have been developed that make algebraic ideas accessible to all students. A traditionally difficult goal in algebra is helping students make the transition from the explicit nature of ideas and symbols in arithmetic to their multiple meanings in algebra. One innovative approach is to use table building to illustrate how algebraic symbols can be used to generalize a relationship.

Complete the following table for rectangles whose length is three times their width.

Width	Length	Perimeter	Area
4	3(4)	$2 \cdot 4 + 2 \cdot 3(4) = 32$	$3(4) \cdot 4 = 48$
5½			
8.2			
w			

When this table is completed, the student's generalization in the last row

w	$3w$	$2 \cdot w + 2 \cdot 3w = 8w$	$3w \cdot w = 3w^2$

can be incorporated into simple programs that use this result as the print statement:

```
10 READ W
20 PRINT W, 3 * W, 8 * W, 3 * W ^ 2
30 DATA 4, 5.5, 8.2

10 FOR W = 4 TO 10 STEP .5
20 PRINT W, 3 * W, 8 * W, 3 * W ^ 2
30 NEXT W
```

This kind of development lets students build on their arithmetical knowledge to create generalizations that when expressed in algebraic form are variable in nature and well suited for performing repetitive tasks.

> Developmental learning approaches to mathematics consider the learner rather than the mathematical content as the variable in achieving curriculum goals. The materials developed according to this principle have been validated with students representing a much wider spectrum of abilities than have the more traditional materials suited to ability tracking. Typically, such materials employ a wide variety of learning activities, evaluation methods, and reinforcement means.

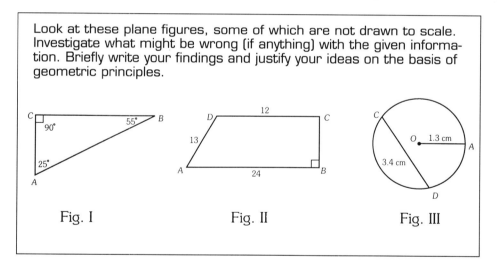

Evaluation and Assessment

The most sensitive areas of the instructional process are evaluation and assessment—the points at which individual philosophies and professional prerogatives grapple with program realities, expectations, and accountability. Teachers are often torn between the merits of major curricular changes and their misgivings over their ultimate lack of control over fundamentals of their programs: *standardized testing, placement of students, college entrance exams, parent demands, administrative constraints, apprehension toward major revisions, collegial indifference.* These legitimate issues are addressed in chapter 5 when we suggest how to implement a core curriculum. However, the climate for aligning evaluation and assessment practices with the curriculum has never been more conducive than at present. Policy developers are increasingly recognizing the power that such practices wield in influencing instruction and student outcomes.

Organizations outside the classroom are restructuring assessment practices to reflect the changing goals of mathematics instruction and the improved ways of certifying a society of mathematics literate citizens. Major innovative projects are under way at educational testing companies to assess higher-level skills and outcomes, multistep problem solving, and uses of technology as a part of assessment. State departments of education are recognizing the needs for multiple forms of assessment and are revamping their assessment procedures. One effort by the California Assessment Program is focusing on students' written responses to the following type of open-ended questions (fig. 1.6).

Look at these plane figures, some of which are not drawn to scale. Investigate what might be wrong (if anything) with the given information. Briefly write your findings and justify your ideas on the basis of geometric principles.

Fig. 1.6

There is a growing use of performance items of this type on state and national assessment instruments. Written responses give evaluators a much clearer idea of students' thinking processes. Questions eliciting open-ended responses require more holistic approaches for scoring. Indirectly, they convey to students the need to communicate their ideas clearly and to construct their responses for a purpose. The impact on the curriculum of this type of assessment is to hold students accountable for *demonstrating* their understanding of connected ideas rather than displaying their proficiency with disconnected skills.

Educational policymakers are beginning to appreciate the differences between individual student assessment and program evaluation and how their uses of such data affect curriculum development and the politics of

There is a bit of nostalgia in each of us longing to preserve what we have experienced and come to value. Changing our behaviors requires us first to modify our views—to believe in the worthiness of the goal.

change. They are becoming sensitive to the impact of assessment on teaching practices. Although the desire for improved student performance will continue to drive the collection and the analyses of large-scale student assessments, there is an increasing understanding of the importance of correlating changes in assessment procedures and objectives with changes in curriculum.

IN SUMMARY

The nature of mathematics education is that it is always changing, subject to new discoveries, to new uses, and to emerging needs for mathematical literacy. We have attempted in this chapter to provide an appreciation for the strength of the forces shaping current changes and the accelerating pace for instituting those reforms. We have not appealed to student achievement results nor to international comparisons as a rationale for change. The needs are far more subtle and far more serious than outperforming an earlier cohort group or winning an international mathematics competition. Before us is a challenge to reshape mathematics education to respond to broad societal changes, to changes in the nature of our discipline, and to changes in our ability to nurture mathematical growth. These factors affect *all* students. By necessity, they require reexamination of the whole curriculum and the assumptions on which it rests.

Figures A through E represent targets for dart throwing.

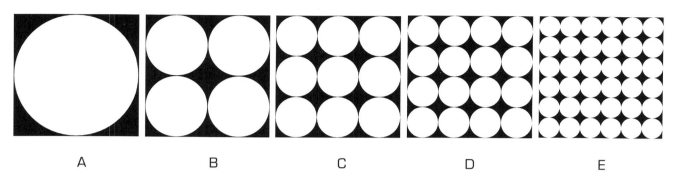

A   B   C   D   E

Suppose that you can earn points by throwing darts according to these rules:

You score 1 point, if your dart lands inside a circle.

You score O points, if your dart lands inside a square but outside a circle.

Darts thrown outside a square do not count (you can throw them again.)

Which target would you choose to throw at? _____

Why?

How can you support your answer mathematically?

Hint 1: Think about the combined areas of circles within a square and the area of that square.

Hint 2: What is a "nice" number to choose for the side length of the squares in order to perform the calculations easily?

Describe your plan in words.

DART THROWING (CONTINUED)

SHEET 2

Information for All Targets

Side length of square: ____________ units

Area of square: ____________ square units

Leave all results below in terms of π.

Target A

Radius of one circle: ____________ units

Area of one circle: ____________ square units

Ratio: ________________________________ = ____________
(state in words) (value)

Target B

Radius of one circle: ____________ units

Area of one circle: ____________ square units

Area of ____________ circles: ____________ square units

Ratio: ________________________________ = ____________
(state in words) (value)

Target C

Target D

Target E

What do you conclude?

Express the area ratio(s) for all targets as a decimal. What fraction or percentage of the time would you expect a dart to land inside a circle if you knew that it hit inside the square?

Experimental Results

Part I Your teacher will supply target A and another target.

Choose a member of your group to be the dart thrower. Measure a distance eight feet from the target and place a piece of tape on the floor. Standing behind this tape, the dart thrower throws some number of times (your choice) at target A. Keep a tally of darts (a) inside the circles and (b) in the square but outside any circle. Darts outside the square do not count.

CAUTION: All members of the group must stay behind the thrower, and only the thrower retrieves the darts.

Repeat this activity with another target supplied by your teacher.

For each target, calculate as a decimal the ratio of the darts inside the circles to the total darts hitting the target.

Target A   Target _____

$$\frac{darts\ inside\ circles}{darts\ hitting\ target} =$$ $$\frac{darts\ inside\ circles}{darts\ hitting\ target} =$$

ratio: ratio:

Are these ratios the same? Are they about the same as your "mathematical" results on sheet 2? What do you conclude?

What decision(s) did you have to make as you did the activity? Might they have influenced your ratios? Why or why not?

Experimental Results

Part II   You will use the same two targets for this activity that you used for part I.

Choose another member of your group to be the dart thrower. Measure a distance six feet from the target and place a piece of tape on the floor. Standing behind this tape, the dart thrower *with eyes closed* throws some number of times (your choice) at target A. Counting only darts that hit the target, keep a tally of darts (a) inside the circles and (b) outside the circles.

CAUTION: All members of the group must stay behind the thrower, and only the thrower retrieves the darts.

Repeat this activity with another target supplied by your teacher.

For each target, calculate as a decimal the ratio of the darts inside the circles to the total darts hitting the target.

Target A	Target _____
$\dfrac{darts\ inside\ circles}{darts\ hitting\ target} =$	$\dfrac{darts\ inside\ circles}{darts\ hitting\ target} =$
ratio:	ratio:

Are these ratios the same? Are they about the same as those computed in part I? Are they about the same as your "mathematical" results on sheet 2? What do you conclude?

CHAPTER 2
MAKING MATRICES ACCESSIBLE TO ALL:
THE DUTCH PERSPECTIVE

There is nothing more difficult to take in hand, more perilous to conduct, or more uncertain in its success, than to take the lead in the introduction of a new order of things.

—Machiavelli

INTRODUCTION

The design of a core curriculum must encompass more than a reorganization of students and topics. We will need to think deeply about mathematics content, its uses, and its potential significance to students as well as about our approaches to teaching. An outgrowth of this consideration is that we may need to develop some fresh perspectives on mathematics topics and their contemporary applications.

For example, the topic of matrices is traditionally treated in North American curricula for college-intending students in formal and usually limited ways in grades 11 or 12. In contrast, our counterparts in the Netherlands view this topic as a powerful, meaningful, unifying concept accessible to all secondary students through many interesting contexts. In this chapter, we take a closer look at matrices as an exemplar of a core topic that can be made meaningful to broad and diverse student populations.

Matrices have been in the Dutch curriculum for all students since 1985. Matrix topics and applications are currently introduced in grades 8 and 9. The Dutch see matrices as an excellent vehicle for addressing several major goals: problem solving, modeling and its limitations, interpreting results, and recognizing isomorphisms. Moreover, they have found matrices to be a rich area for student investigations and one that allows differentiation of learning outcomes.

The examples that follow were first introduced in the Netherlands in 1981 and have since been revised and piloted in the United States with ninth-grade students at Whitnall High School (Wisconsin) in 1990 (de Lange 1990). They should help convey the spirit of the core curriculum—a practical, interesting, accessible approach to sound mathematics for all secondary students. This discourse has been prepared by Jan de Lange of Utrecht University, the Netherlands, and includes a description of, and examples from, *Matrices*, the booklet used at Whitnall High School.

THE BOOKLET MATRICES

Developing each student's ability to solve problems is essential if he or she is to be a productive citizen. As the *Agenda for Action* (NCTM 1980, p. 2) states:

> Problem solving must be the focus of school mathematics.

The *Curriculum and Evaluation Standards for School Mathematics* (NCTM 1989) adds that to develop such abilities, students need to work on problems that may take hours, days, or even weeks to solve. Some of these problems should be relatively simple exercises that can be accom-

Assessment Matters:

Changes in our view of teaching and learning mathematics as described in chapter 1 and illustrated here require similar changes in how we assess student performance. Most traditional tests focus on retrieval of information and mastery of procedures. Students soon learn that the reward system defined by these tests requires diligent memorization and practice. Activities that are designed to help students construct a deeper understanding of mathematical concepts are seen as having little payoff, at least little immediate payoff. Some alternative methods of assessment are identified toward the end of this chapter.

plished independently; others should involve small groups or an entire class working cooperatively. Some problems should be open-ended with no right answer; others should require students to formulate and answer questions corresponding to given situations.

Problem solving is often seen as something that can be done only after mathematical skills have been mastered. This is not the case in the materials that we have been developing in the Netherlands for the past fifteen years. Problems form the start for every learning cycle, and mathematical concepts are developed or constructed along the way. In the classroom, this means that the teacher does *not* begin the lesson with an explanation but that students begin working together on a problem from their student text without interference from the teacher.

Problem solving often requires integrated and interrelated mathematics. In real problems, we cannot always stick within the boundaries of algebra or geometry. So, it will come as no surprise that the booklet on matrices starts with many problems from different areas of mathematics.

Basically, the first chapter is about *distances.* In mathematics, distances are precise and can be computed easily—the distance between two points whose coordinates are given or the length of the side of a triangle. In real life, problems are not that simple. On the following part of a map, we see the driving distances between selected cities. The map also implicitly gives the distances "as the crow flies."

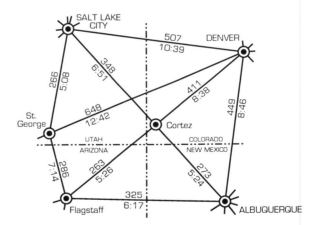

We ask students to calculate the distances as the crow flies by giving them the scale of the map. A number of other questions can now be asked. For example:

◆ What is the shortest route from Salt Lake City to Albuquerque in miles or kilometers?

◆ What is the fastest route in hours and minutes?

◆ Compare the road and air distances. Between which two cities is the "car distance" much longer than the "airplane distance"?

Progressing from the map to a distance table or a matrix is a natural step. From any given map we can always make a distance table. But for most students the following problem is a real surprise:

Use your ruler and compass to draw a map that shows the correct distances between points A, R, U, as given by the following distance table:

$$\text{from}$$

$$\begin{array}{c} \\ \text{to} \end{array} \begin{array}{c} \\ A \\ R \\ U \end{array} \begin{array}{ccc} A & R & U \\ \left[\begin{array}{ccc} 0 & 70 & 50 \\ 70 & 0 & 60 \\ 50 & 60 & 0 \end{array}\right] \end{array}$$

For us as observers, there was another surprise: almost none of the students had a compass and a ruler. We found out that those mathematical tools were used very seldom and were used exclusively in geometry lessons. And this was algebra, wasn't it? The following two problems added to the students' state of surprise:

Draw a map showing the correct distances to correspond with the following distance table. What could be the approximate values for x?

$$\text{from}$$

$$\begin{array}{c} \\ \text{to} \end{array} \begin{array}{c} A \\ R \\ U \\ L \end{array} \begin{array}{cccc} A & R & U & L \\ \left[\begin{array}{cccc} 0 & 70 & 50 & 50 \\ 70 & 0 & 60 & 40 \\ 50 & 60 & 0 & x \\ 50 & 40 & x & 0 \end{array}\right] \end{array}$$

Try to draw a map showing the correct distances to correspond with the following distance table:

$$\text{from}$$

$$\begin{array}{c} \\ \text{to} \end{array} \begin{array}{c} P \\ Q \\ R \end{array} \begin{array}{ccc} P & Q & R \\ \left[\begin{array}{ccc} 0 & 10 & 50 \\ 10 & 0 & 30 \\ 50 & 30 & 0 \end{array}\right] \end{array}$$

We wanted to explore the concept of distance a little further. So social distances, differences in excavations, and similarities in pottery were discussed. Try to imagine for yourself the nature of teacher and student interaction as the following tasks from the booklet were used.

Excavations

While excavating the lost Mayan civilizations in Central America, the archaeologist Robinson decided to notate the "distance" or "measure of difference" between the various excavations in the following way. He first divided the excavated remains into different classes:

Class 1: human bones
Class 2: animal bones
Class 3: pottery
Class 4: refuse matter
Class 5: clothing

Robinson then indicated with percentages the distribution of the five classes in each excavation:

	Class				
	1	2	3	4	5
A	10	25	35	20	10
B	0	40	30	25	5
Excavation C	5	10	25	40	20
D	25	25	25	15	10
E	0	10	25	40	25

- In which excavation(s) did he find no human bones?
- Which two excavations somewhat resemble each other, according to the table?

Robinson took the following distance measure among the five excavations; let's look, for example, at the distance or degree of difference between A and B:

	1	2	3	4	5
Excavation A	10	25	35	20	10
Excavation B	0	40	30	25	5
Distance	10	15	5	5	5

The total distance between A and B can be expressed as follows:

$$d(A, B) = 10 + 15 + 5 + 5 + 5 = 40$$

Assessment Matters: **Coop-erative group work provides an opportunity for informal, individual assessment that is not easily obtained during direct instruction or seat work. As the groups work and discuss, the teacher can circulate around the room, unobtrusively observing the level of participation of individual students and how well they apply key concepts and problem-solving strategies. It is a good idea to record these observations on a simple checklist that includes space for short comments. Focus on a few students each day, with a goal to observe carefully each student once every two weeks.**

Complete this distance table indicating the distances among the five excavations:

	A	B	C	D	E
A	0				
B		0			
C			0		
D				0	
E					0

- Which excavations are "far apart"?
- Which excavations are "close together"?
- How might such a table be useful for archaeologists?

Characteristics

Pieces of a number of vases were found during the excavations. These vases showed the following characteristics:

with or without a handle

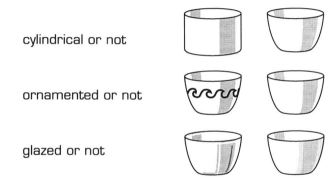

cylindrical or not

ornamented or not

glazed or not

Six vases were examined with regard to these characteristics:

	han*	cyl**	orn+	glaz++
Vase a	1	0	1	0
Vase b	0	1	0	0
Vase c	1	0	1	1
Vase d	0	1	0	1
Vase e	0	1	1	1
Vase f	0	0	0	1

```
 *   1: handle          0: no handle
 **  1: cylindrical     0: not cylindrical
 +   1: ornamented      0: not ornamented
 ++  1: glazed          0: not glazed
```

◆ Which cylindrical vase has no handle, is not ornamented, but is glazed?

Here, too, you can easily draw a degree of difference table. See how many of the four characteristics the vases have in common. The distance between vase a and vase b, for example, is 3: their only similarity is that neither is glazed. So they differ on three characteristics.

◆ Try to find vases that are "close together," in other words, that only differ on one characteristic.

◆ Draw a distance table for the six vases. For distance, use degree of difference as described above.

———————————————

The vase example offers many opportunities for further exploration.

The next two chapters of the *Matrices* booklet treat connectivity matrices and connectivity graphs, similar to the example given in the standard on discrete mathematics (NCTM 1989, p. 177). Other distances are discussed once more in the fourth chapter where Pythagorean distance and taxi distance are compared. An example of a straightforward exercise from this chapter follows:

———————————————

◆ ◆ ◆ ◆ ◆ ◆ ◆ ◆

Make a distance table for the three points A, B, and C on the grid below by using—

a. Pythagorean distance

b. taxi distance

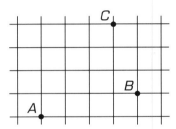

The first four chapters are concerned only with matrices as representation tools. There is no addition, subtraction, or multiplication of matrices. Students write matrices from graphs or maps and have to interpret and compare them. Matrices and graphs, maps and distances are all interrelated. There is not always one correct answer, and a lot of class discussion and interaction takes place.

The second part of the booklet concerns the more dynamic part of matrices. Students are confronted with a production problem.

A certain factory produces simple furniture—bookcases and tables. In fact, they make only four basic products or elements: (1) a table top; (2) a table leg; (3) a bookshelf; and (4) a vertical shelf support. These four elements are sold separately, but they are also combined and sold as finished products: a table, a bookcase, a bookcase element (consisting of a bookshelf and two vertical supports), and a table-bookcase combination. The eight products are shown below.

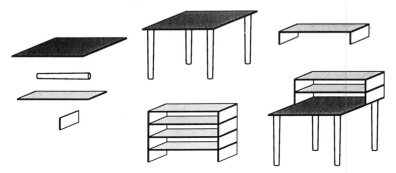

The production schema indicates that to produce a bookcase, (1) eight vertical supports and four shelves are produced; (2) four complete elements are assembled; and finally one bookcase is produced.

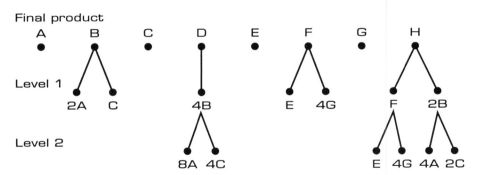

For another factory, we have the following production graph:

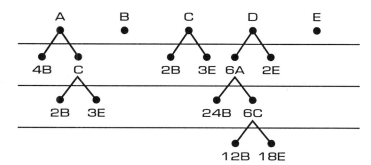

This, in turn, leads to the following production matrices (this discussion and presentation takes several pages in the booklet).

Matrices

Level 0

Level 1

Level 2

Level 3

Total requisite

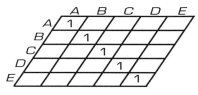

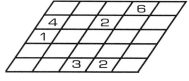

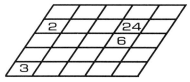

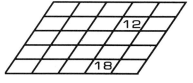

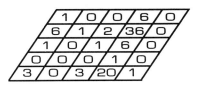

At the end of the *Matrices* booklet we come back to this problem with the following questions.

♦ We will call the first matrix L_0, the second one L. Calculate $L \cdot L$. Which matrix is the same as $L \cdot L$? Explain.
♦ Calculate $L \cdot L \cdot L$. Which matrix is the same as $L \cdot L \cdot L$? Explain.
♦ Suppose $L_0 = L^0$. Calculate $L^0 + L^1 + L^2 + L^3$. What is the significance of this matrix?

These questions are not easy to answer. The isomorphism between the different problems becomes even more clear if we consider the following problem.

The Boss

Graphs are also used to determine the relationships within groups. One way of doing this is to take different pairs from a large group and see which member of each pair is the "boss" over the other. We're looking for the leader of the large group. A graph that illustrates this kind of relationship is called a domination graph. It is an example of a directed graph:

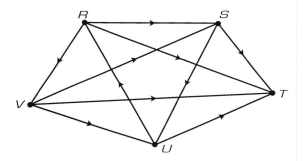

♦ Determine the accompanying domination matrix, D (a matrix whose i–j entry is 1 or 0, depending on whether there is a directed edge from i to j).
♦ Who is the leader of the group? Who can't boss anyone at all?

After studying the graph or matrix, we can see that the answer to the leadership question is that there is no one leader. R and V (see the number of ones in the corresponding rows of matrix D) can both justifiably be called the leader.

To overcome this deadlock, we can look at the two-stage domination.

♦ Determine D^2. What is the meaning of D^2?
♦ Determine D^3. What is the meaning of D^3?
♦ Determine $D^0 + D^1 + D^2 + D^3$ and explain what it means.

It goes without saying that multiplying matrices has been introduced along the way. Space considerations do not allow discussion here of that part of the booklet, but we may state that students discover matrix multiplication by themselves in a very natural way. Interesting contexts from which students can construct meanings for matrix operations may be found in *Connecting Mathematics* (Froelich et al. 1991), another volume in the Grades 9–12 Addenda series.

Below are two examples from the *Matrices* booklet that illustrate our perspective on matrix multiplication.

Jukebox

A dealer in old jukeboxes keeps track of his merchandise in a modern way—with a computer. His computer holds information concerning his stock, the number of sales each week and each month, and, of course, his prices and profits.

In April he sold the following jukeboxes:

$$
\text{date of manufacture}
\begin{array}{c}
 \\ 1958 \\ 1960 \\ 1962
\end{array}
\begin{array}{c}
\text{brand: Wurlitzer} \quad \text{Rock-Ola} \quad \text{Seeburg} \\
\left[\begin{array}{ccc}
2 & 1 & 0 \\
0 & 3 & 1 \\
1 & 2 & 3
\end{array}\right]
\end{array} = S \,(\text{elling matrix})
$$

The average cost-price and profit expressed in a matrix (in dollars) was as follows:

$$
\begin{array}{c}
\text{Wurlitzer} \\ \text{Rock-Ola} \\ \text{Seeburg}
\end{array}
\begin{array}{c}
\text{Cost-price} \quad \text{Profit} \\
\left[\begin{array}{cc}
5000 & 2000 \\
3500 & 1500 \\
3000 & 1500
\end{array}\right]
\end{array} = D
$$

♦ Calculate $S \cdot D$ and explain the meaning of this matrix product.

In other cases, the meaning of a matrix product may be less obvious. The following problem from the *Matrices* booklet gave our project teachers (in their short course before they did the actual teaching) quite some thought.

Concerts

A concert organization office has booked four pop groups who will perform in four cities according to the following matrix:

$$
\begin{array}{c}
\text{Amsterdam} \\ \text{Utrecht} \\ \text{Rotterdam} \\ \text{Eindhoven}
\end{array}
\begin{array}{c}
U_1 \quad U_2 \quad U_3 \quad U_4 \\
\left[\begin{array}{cccc}
2 & 3 & 0 & 1 \\
0 & 2 & 2 & 3 \\
1 & 1 & 1 & 2 \\
1 & 0 & 2 & 1
\end{array}\right]
\end{array} = S
$$

Teaching Matters: During this period of transition to a core curriculum, teachers will need to draw and experiment with ideas from many sources. Other resources for incorporating matrices in secondary school mathematics include the 1988 NCTM booklet Matrices *by the Department of Mathematics and Computer Science of the North Carolina School of Science and Mathematics, the 1991 NCTM Yearbook,* Discrete Mathematics across the Curriculum, K–12, *and* Geometry from Multiple Perspectives *(Coxford et al. 1991).*

146.516

The tickets are available in three price categories and also differ for each pop group:

$$\begin{array}{c} \quad \text{Category} \\ \quad 1 \quad\;\; 2 \quad\;\; 3 \\ \begin{array}{c} U_1 \\ U_2 \\ U_3 \\ U_4 \end{array} \left[\begin{array}{ccc} 45 & 30 & 25 \\ 50 & 30 & 30 \\ 35 & 25 & 20 \\ 25 & 20 & 15 \end{array} \right] = P \end{array}$$

Here are the number of people in the audience at each concert (rounded off):

$$\begin{array}{c} \quad U_1 \quad\;\; U_2 \quad\;\; U_3 \quad\;\; U_4 \\ \begin{array}{c} A \\ U \\ R \\ E \end{array} \left[\begin{array}{cccc} 2100 & 900 & 0 & 900 \\ 0 & 2100 & 1500 & 1500 \\ 1500 & 900 & 6000 & 1200 \\ 900 & 0 & 1500 & 900 \end{array} \right] = B \end{array}$$

Roughly one-sixth of the tickets sold were in the first category, one-third in the second category, and one-half in the third category.

♦ Which pop group had the largest total audience?

♦ Complete the following matrix:

$$\begin{array}{c} \text{Amsterdam:} \quad U_1 \;\; U_2 \;\; U_3 \;\; U_4 \\ A = \begin{array}{c} 1 \\ 2 \\ 3 \end{array} \left[\begin{array}{cccc} \cdot & \cdot & \cdot & \cdot \\ \cdot & \cdot & \cdot & \cdot \\ \cdot & \cdot & \cdot & \cdot \end{array} \right] \end{array}$$

♦ Calculate $P \cdot A$.

♦ What does this matrix product mean?

♦ Can $S \cdot P$ be calculated?

♦ Does $S \cdot P$ mean anything?

As can be seen from the previous examples, our principles include the emergence of mathematical knowledge from problem situations. The mathematical knowledge at the very basic skill and concept levels includes defining a matrix, matrix multiplication by a scalar, matrix multiplication, powers of matrices, graphs, directed graphs, maps, and distances in very diverse contexts. At least equally important are these goals: modeling, interpreting solutions, argumentation, creativity, communication, interaction, and discussion. Experiencing real problems, with different solutions that depend on the judgment of the student, should be a part of the core curriculum.

ASSESSMENT

During experiments that eventually led to new curricula in the Netherlands, we were confronted with a serious problem—timed written tests were inappropriate for proper assessment of our intended learning outcomes. So we started our developmental research into new test formats. We followed five principles:

1. Tests should be an integral part of the learning process, so that tests improve learning.
2. Tests should enable students to show what they know, rather than what they do not know (i.e., positive testing).
3. Tests should operationalize both lower- and higher-order thinking skills.
4. The quality and format of the test should not be dictated primarily by its possibilities for objective scoring.
5. Tests should be practical to administer.

We have given numerous examples of what we believe to be appropriate test items in a number of publications, including the Grades 9–12 Addenda series book, *Data Analysis and Statistics across the Curriculum* (Burrill 1992). The test formats include the following:

- Restricted-time written tests
- Two-stage tests
- Take-home tests
- Oral tests
- Student-produced tests

We will expand on one of these, the two-stage test. The two-stage test consists of both open-response and essay questions. The first stage is carried out like a traditional, timed written test. Students are expected to select and answer as many questions as possible within a fixed time limit (e.g., 45 minutes). In principle, students are free to tackle any question they like during this time. However, students usually prefer to answer the open-response questions first and to reserve the essay questions for later.

After having been scored by the teacher, the tests are handed back to the students with major mistakes noted. Now the second stage takes place. With this additional information, the student repeats the work at home without the restrictions of the classroom environment and is completely free to answer the questions as he or she chooses. Students may answer the questions in the order stated, by means of an essay, or as a combination of methods. After a certain time, say three weeks, students turn in their work, and a second scoring takes place. This provides the teacher (and students) with two marks—a first-stage one and a second-stage one. The result is a much more accurate assessment of the students' understanding and capabilities.

CONCLUSION

Matrices are an example of how a powerful and rich area of mathematics can be extended to the mainstream preparation of all students. The challenge is to free our thinking from traditional dichotomies about content and students. We must recognize that even our most ambitious thinking is limited by our past experience. The Dutch examples of matrix applications to real-world situations and the connection of ideas in applying mathematics without concern for "course packages" are both illuminating and sobering—illuminating because they reflect applications that we may not have seen and sobering because we may have to come up to speed on teaching some mathematics quite different from that in our own experience or formal training.

Central to the notion of teaching broadly useful mathematics to a more diverse range of students is the approach of letting the mathematics arise from students' experiences and supporting the students' learning through developmental teaching. It is to these two features that we now direct your attention in the next two chapters on core curriculum models and sample instructional lessons.

CHAPTER 3
A CROSSOVER CURRICULUM MODEL

The focus of school mathematics is shifting from a dualistic mission—minimal mathematics for the majority, advanced mathematics for a few—to a singular focus on a common core...for all students.
—Everybody Counts

This chapter and the next provide examples of curriculum models and abbreviated prototype lessons that illustrate how a core program might be structured and implemented in classrooms. Three models, *Crossover, Enrichment,* and *Differentiated,* are discussed in these chapters. Generally, the models represent a continuum of transition stages with the Crossover Model closest in structural similarity to traditional curricula. Sample syllabi keyed to the Crossover and Differentiated Models may be found in Appendixes I and II respectively.

Each curriculum model is briefly described according to the philosophy of its approach and the implications for its use. Following the model description are a number of illustrative classroom lessons that refer to the appendix syllabus for that model. Each of these lessons focuses on a significant mathematical topic and would extend over two or more class sessions. Each lesson is introduced through a question to motivate students and set the stage for the learning activities to follow. Through activities and follow-up class discussion, learners are expected to *construct* their own understanding of the major ideas of that lesson.

The lessons are activity based, with the students engaged in conjecturing, inventing, reasoning, and problem solving. It is intended that the teacher will serve as a guide and a facilitator in helping students to connect the lesson ideas and applications. Each lesson addresses the mathematical ideas at four or more levels of work as described in the *Curriculum and Evaluation Standards* (NCTM 1989, pp. 131–36). Sometimes extensions of the initial lesson ideas make up successive levels. At other times, differentiated activities are provided for students ready to engage concepts at different levels of understanding and abstraction. There are some differences in the way that lessons are structured. These differences reflect the contrasting perspectives of the authors on how such lessons could be developed. Differences have been retained to give the reader a range of possible approaches for teaching core topics.

The Crossover Model

The Crossover Model (fig. 3.1) consists of two parallel course sequences, A and B, that each follow the same syllabus (see Appendix I). The first three years of the syllabus constitute the core. The fourth year addresses additional topics for college-intending students. The Crossover Model is considered a transition model that will enable school districts to address core outcomes for all students within existing curriculum structures. The two parallel course sequences permit grouping students by learning needs and career aspirations concerning mathematics. Topics continue to be arranged by courses that correspond approximately to the existing algebra-geometry-advanced algebra-precalculus sequence. These choices have been made to offer a model that contains features of existing course alignments.

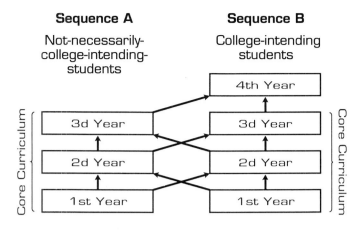

Sequence A
Not-necessarily-college-intending-students

Sequence B
College-intending students

Any student may progress to the next higher course vertically or diagonally with an equivalent core coverage of prerequisite topics.

Fig. 3.1. Crossover Curriculum Model. Two parallel strands with equivalent core outcomes at the same horizontal levels

*Teaching Matters: **Movement from one sequence to the other may not always be smooth. However, it need not be more difficult than moving from, say, algebra to geometry to a second year of algebra in traditional sequences. Adaptive teaching that takes into account students' thinking can make a given course a successful experience for students even if their mathematical backgrounds differ significantly.***

Features of the model

Several curriculum features built into this model address the scope and philosophy of the *Curriculum and Evaluation Standards.* The first key feature is the design of a new sequence of core courses for students who traditionally have shunned or have been tracked out of college-preparatory courses. The A-course sequence (for not-necessarily-college-intending students) strongly emphasizes developmental methods for building concepts through a variety of contexts derived from real-world situations. Learning centers on concrete activities and models, with less emphasis given to formalism, computational or symbol-manipulating facility, and mathematical structure. The B-course sequence moves quickly, if at all, through similar concrete activities and then continues on to explorations of more sophisticated generalizations. The requisite time for a heavier emphasis on developmental instruction in the A-course sequence (compared to the B-course sequence) is achieved by reducing the time devoted to more complex exercises, enrichment topics, and connections with advanced mathematical ideas and applications.

The second key feature of the model is parallel course sequencing of the same core concepts. Consequently, students in each sequence are learning the prerequisite ideas, within the same scope and sequence, for entering the next course of *either sequence,* should there be a change in student career interests, maturity, or learning needs. The opportunity to *cross over* from the A sequence to the B sequence or the reverse allows achieving the core curriculum goals in three years with considerable flexibility. Both sequences carry comparable academic integrity, keep career opportunities open, and hold comparable value for achieving program exit goals for student competence.

The third key feature of the model is the altering of topics within courses to promote a *constructive* approach to mathematics within each sequence. For example, syllabus headings for the first three years carry conventional headings in quotations—"algebra," "geometry," and "advanced algebra." But topics within a given course appear in an order more conducive to developmental activities and building connections among ideas. New core topics appear in descriptions for a particular "course," and some topics are delayed in the syllabus or de-emphasized from traditional course syllabi. For example, some geometry topics appear in "algebra" and "advanced algebra" and some algebra topics appear in "geometry." Statistics and matrices appear in each course. Manipulative facility with complex expressions is de-emphasized throughout.

Curricula organized around familiar content packages like "algebra" provide less fertile ground for developing connections, reasoning, and problem solving than a fully integrated curriculum does. But this more traditional approach furnishes a manageable and realistic *transition curriculum* for teachers adjusting to new expectations for themselves and their students. The Crossover Model syllabus is also more likely to coincide with the current availability or adaptation of materials.

Instructional Considerations

The Crossover Model adheres strongly to the principle that students must construct their own ideas and build understanding through connections. This view determines not only the way that instruction is organized by lessons but also the way in which topics are ordered for study. For example, in the "algebra" course, the notion of linear relationships is introduced informally after a statistics chapter as a way of predicting rules for well-behaved, paired data. Intuitive notions of slope and intercept are introduced through observations about families of lines. But the slope-intercept form of a linear equation is not introduced until the last chapter, after students have had multiple experiences with graphing, associating the elements of ordered pairs, expressing variation relationships, and writing formulas to express relationships from written, tabular, and graphical data.

A developmental approach makes ideas much more accessible to more diverse groups of students, de-emphasizes manipulative facility, provides a richer mixture of connected ideas and applications, and delays the formalism of mathematics. The two parallel sequences of the Crossover Model make no assumptions about the ability of students who elect to begin (or move to) the A or the B sequence. The terms *not-necessarily-college-intending* and *college-intending* can be somewhat misleading for the two sequences. All students are considered potential candidates for post–high school education or training and require the core as a foundation for their mathematics preparation. The fact that both sequences are expected to accomplish a common set of core understandings should make the task of large-group assessment easier. All students can be assessed over the same sets of goals, a particular advantage as assessment techniques incorporate more performance features and open-ended measures of understanding.

What is not apparent from the syllabus is the expectation that all students will take full advantage of group work and activities that engage them in reasoning, communication, and problem solving. Students in both sequences, but particularly the A sequence, are expected to engage in guided discovery activities and in work with physical models that actively involve them in thinking and sharing their learning and ideas. In the A sequence, the teacher will likely play a somewhat more active role in structuring learning into manageable, developmental increments, with lessons and learning activities frequently centering on full-period interactive instruction.

This model assumes some access to technology but does not depend heavily on it, particularly at the level of the first two courses. This is a pragmatic consideration of the model. Local financial constraints, as well as "people changes," determine how much can be accomplished immediately. In the initial stages of implementing this model, districts may place more emphasis on obtaining the physical models and teacher-made materials necessary for the developmental approaches. The third and fourth courses do assume the availability of graphing utilities. Much of the work in both the A and B alternatives involves exploratory activities in introducing functions and determining the effects of parameter changes on graphs.

The descriptions for the first two courses in both the A and the B sequences contain the same listing of topics, although the treatment of these topics and the approaches may vary considerably. This is necessary for the "crossover" options of the model and the prerequisite core understandings needed to move to the next course in either sequence. Course 3 represents a slightly different situation. The students opting to go on to the fourth year of the curriculum are expected to be college intending. The third year in the B sequence will be the strongest preparation for that fourth year. In the appendix syllabus for the third year, topics preceded with the (Δ) icon at the end of each chapter listing are intended only for the B course. All other topics in the syllabus are intended for both courses.

The Challenge of the Model

The full three- or four-year listings of topics for this model may appear robust, particularly for all students. One should be reminded, however, that each year of the sequence is built on the *understandings* developed in the preceding year. The emphasis within the core is primarily on developing understanding rather than manipulative facility. Moreover, every topic within this curriculum has been selected because it is a *core* idea necessary for intelligent citizenship and for a sound mathematical foundation for future work and study.

Explanation of the Lesson Formats

Each lesson—covering two or more days of instruction—is identified by a title describing the major ideas. A *location note* indicates where each lesson occurs in the syllabus so that the reader can scan the topics covered prior to this unit; prerequisite ideas for the lesson are noted as well. The objectives and the major standards addressed are listed for quick reference. Instruction begins with a motivating question that sets the stage for the learning activities to follow, engaging the class in group discussion and conjecturing about the solution of the real-world application. The motivating question would be answered before closing the lesson.

Learning activities for students are then described by levels. Generally, levels 1 and 2 are considered appropriate for A-sequence students and levels 3 and 4 for B-sequence students. Often, there will be an indication that B-sequence students should cover some of the ideas or activities from levels 1 or 2, but more rapidly and in less detail, before they move to level 3 and 4 work. Interspersed are *Class Discussion* sections that list questions to guide students in focusing on the major ideas of the activities or in consolidating their understanding. Periodically, there are *Teaching Matters* notes in the sidebar, which elaborate on the treatment of the activity or which furnish extensions that might be appropriate. Answers to selected exercises appear in brackets.

INTUITIVE LINE FITTING AND INTERPRETING LINEAR GRAPHS

Location in sample syllabus: Year 1, Unit 3

Major standards addressed: Statistics, algebra, functions

Objectives: ◆ To recognize the relationship between an equation of a line and the coordinates of points on the line

◆ To develop intuitive notions about the general locations of graphs of the form $y = ax$ and $y = x + b$

Prerequisites: Introduction to the coordinate system, plotting ordered pairs, and vocabulary—axes, coordinates, quadrant

Sequence A will address level 1, and possibly level 2, of the lesson activities. Sequence B will provide some exploratory work similar to that in levels 1 and 2, but will focus on activities at levels 3 and 4.

Motivating question: A class conjectured that a possible measure of a person's general physical condition was a comparison of a person's height to that person's waist size. To get an idea whether there appeared to be a good relationship between these two measures, the class plotted waist size versus height for a random sample of ten students supplied by the physical education department. They then wondered if they could express the relationship as a formula, use that formula to make predictions about other people their age, and successfully apply such "formula" predictions to other age groups.

Level 1

Directions for the initial activities:

The diagram shows some lines that pass through some given data pairs.

1. Why do the data pairs (1, 1), (3, 3), (4, 4) lie on the line $y = x$?

2. Why does the data pair (4, 1) lie on the line $y = \frac{1}{4}x$? Name another data pair that would lie on the line $y = \frac{1}{4}x$.

3. Describe where the graphs of the following lines would appear on the diagram shown:

   a. $y = 2x$ c. $y = 100x$

   b. $y = \frac{1}{10}x$ d. $y = \frac{1}{3}x$

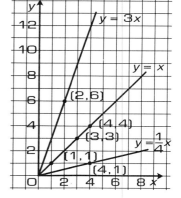

Graph the following data pairs on the grids shown. Draw the line that passes through the pairs and write the equation for that line.

4. a. (4, 8), (2, 4), (3.5, 7)   [$y = 2x$]

   b. (9, 3), (4.5, 1.5), (6, 2)   [$y = \frac{1}{3}x$]

5. a. (10, 5), (2, 1), (6, 3)   [$y = \frac{1}{2}x$]

   b. (0.5, 2), (3, 12), (1.5, 6)   [$y = 4x$]

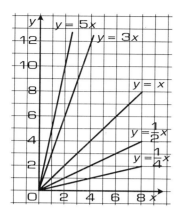

[Diagram for Exercise 4]

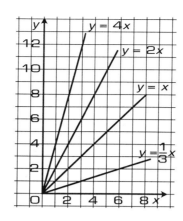

[Diagram for Exercise 5]

Examine the lines and their equations in the diagram below.

6. How are the lines similar? How are their equations similar?

7. What does the numerical term in the equation seem to say about the graph of the line?

8. Graph the following data pairs on the same pair of axes and write the equation for each set of data pairs.

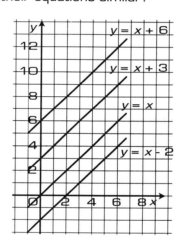

   a. (5, 8), (2, 5), (6, 9)
      [$y = x + 3$]

   b. (1, 2), (9, 10), (4, 5)
      [$y = x + 1$]

   c. (3, 7.5), (1.5, 6), (0, 4.5)
      [$y = x + 4.5$]

   d. (4, 2), (10, 8), (8, 6)
      [$y = x - 2$]

The following equations relate pairs of data. On the same pair of axes, first sketch and label the graph for each equation, then find three data pairs that satisfy each equation and plot and label these pairs.

9. a. $y = 3x$
   b. $y = x + 6$

10. a. $y = x - 5$
    b. $y = 6x$

Level 2

In addition to completing level 1 activities, some students will complete the following activity.

Directions:

a. Explain to students that paired data from real-world activities seldom "fit exactly" on the same line. Rather, points scatter and form a scatter diagram. But if a relationship exists between the two numbers of the data pairs, there may be a line that the points cluster about, with nearly as many points above the line as below it.

Teaching Matters: The focus of these activities is to introduce the equations of linear graphs of the form y = ax and y = x + b, a > 0, as ways of describing relationships among pairs of data. Plotted data pairs arise first, followed by the equation that relates the coordinate values. This more natural development illustrates how algebra is applied to describe patterns and relationships that exist in the real world. Once that relationship is described algebraically, the equation can be used to predict additional coordinate pairs satisfying the same pattern. The concepts of slope and y-intercept are developed intuitively as parameters that relate families of lines passing through the origin or that are parallel to the line y = x. Some students may think erroneously that the a-value "rotates" the line y = x rather than stretching the graph of that line. This impression can be remedied by examining the graphs of y = x², y = 2x², and y = ¹/2x².

In exercises 4 and 5, students graph three data pairs on a diagram that already contains other lines and their equations as a hint for the rule that they are expected to determine. Exercises 9 and 10 indicate whether students can determine the graph from the equation. The students can then identify data pairs that lie on each graph by examining the graph or by using the equation and creating tables of values. In exercise 9, students can be asked to find the data pair that lies on each line and then to discuss why either equation seems to describe the rule relating its values.

Assessment Matters: If the instruction is differentiated, the assessment must also be differentiated. One approach is to incorporate more student self-assessment. Ask students to write a brief self-assessment after they have completed a written assignment. Writing in a journal is also a good way to get students to reflect on their own performance.

b. Give students copies of two templates of families of lines on centimeter graph paper.

c. Have students create a scatter diagram on centimeter graph paper for exercises 11–16.

d. Then have students overlay one of the line templates on their scatter diagram to determine the equation of the line that best fits their data.

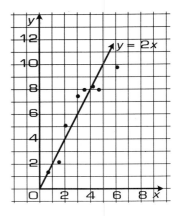

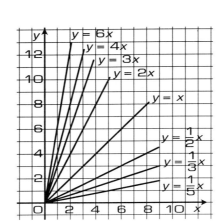

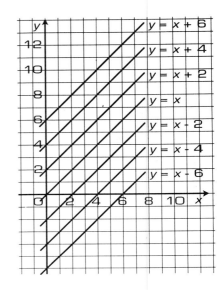

[templates]

11. (4, 7), (0.5, 1), (6, 13), (2, 5), (3.5, 7), (5, 8) $[y = 2x]$

12. (9, 12), (2.5, 8), (4, 8), (6, 11), (1, 5), (7.5, 9.5) $[y = x + 4]$

13.

x	y
0	0.5
2	1
6	2
8	2.5
10	3.5
13	4

$[y = \frac{1}{3}x]$

14.

x	y
2	–4
5	–1.5
7	–1.5
8.5	2.5
10	5
12	5.5

$[y = x - 7]$

15.

x	1	2	3.5	4.5	6	7
y	3	4.5	7	7.5	8	8.5

$[y = x - 2]$

16.

x	0	1	2.5	3	4.5	6
y	0	2.5	8	9.5	12	18

$[y = 3x]$

Level 3

Use a graphing utility to explore exercises 17–21 with a partner.

17. Graph the line $y = x$. Trace with your finger where you expect the line $y = -x$ to appear. Check by graphing the equation.

18. On a set of coordinate axes on your paper, quickly sketch where you expect the graphs of the following lines to be. Check by graphing with your utility or calculator.

 a. $y = -3x$ b. $y = -5x$ c. $y = -\frac{1}{2}x$

19. How are the graphs of $y = 2x$ and $y = -2x$ related? The graphs of $y = \frac{1}{3}x$ and $y = -\frac{1}{3}x$? [Reflections in the y-axis] (Hint: What would happen if you placed a mirror on the y-axis?)

20. How are the graphs of $y = x + 5$ and $y = -x + 5$ related? The graphs of $y = -x - 2$ and $y = x - 2$? [Reflections in the y-axis with y-intercepts of 5 and –2, respectively]

21. About where will the graphs of $y = -100x$ and $y = -\frac{1}{100}x$ be located? Guess first and then check. Be careful!

22. Graph the following data pairs on the same pair of axes and write the equation for each set of data pairs.

 a. $(-4, 4)$, $(-1, 1)$, $(2, -2)$ $[y = -x]$
 b. $(0, 0)$, $(-3, 6)$, $(-2, 4)$ $[y = -2x]$
 c. $(4, -2)$, $(2, -1)$, $(-2, 1)$ $[y = -\frac{1}{2}x]$
 d. $(-2, 3)$, $(-1, 2)$, $(0, 1)$ $[y = -x + 1]$
 e. $(1, 3)$, $(3, 1)$, $(0, 4)$ $[y = -x + 4]$

Level 4

Some Sequence B students will be able to do the following activity.

Another way of thinking about data pairs is a *mapping*. $3 \to 6$, $4 \to 8$, $6 \to 12$ show the same relationship as $(3, 6)$, $(4, 8)$, $(6, 12)$, described more generally by the equation $y = 2x$. Graphically, we can see how the line $y = 2x$ relates the first coordinate value to the second. Using this idea, we can extend the pattern shown by the data pairs to predict the answers to questions such as $10 \to ?$ or $? \to 7$.

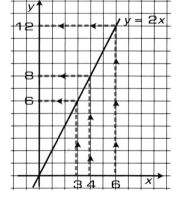

Graph the following data pairs and sketch the line that relates the pairs. Then use the idea of a mapping to determine the missing value from each additional data pair.

23. $(4, -1)$, $(10, 5)$, $(6, 1)$; $(7, ?)$, $(?, 10)$ (Hint: $(7, ?)$ means the same thing as $7 \to ?$)

24. $(12, 3)$, $(6, 1.5)$, $(0, 0)$; $(8, ?)$, $(?, 2)$

25. $(1, 6)$, $(6, 1)$, $(3, 4)$; $(?, 0)$, $(2.5, ?)$

26. $(2, 8)$, $(5, 20)$, $(3, 12)$; $(?, 2)$, $(0, ?)$

Teaching Matters: **In the level 3 activity, students use graphing utilities to explore linear equations of the form $y = ax$ and $y = -x + b$, where $a < 0$. They are expected to discover that lines with negative x-terms are symmetric reflections of their counterpart equations with positive x-terms. A key focus of the activity is to consider the location of all lines as translations of the lines $y = x$ and $y = -x$. In exercise 22, students can be asked to use the graphing utility to plot the points with the given coordinates and to verify that their prediction for the equation of the line passing through the points is correct.**

Teaching Matters: **You might also include in the lesson situations that involve comparing "before" and "after" data with the line $y = x$.**

The graph below gives the cholesterol level for patients before and after a strict diet. What can you say about the effectiveness of the diet in reducing cholesterol level?

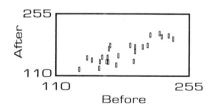

A group of high school juniors is planning to go on a strenuous six-week exercise and strengthening program. How might the exercise program affect the weight of the participants? Will the program have the same effect on overweight students as on underweight students? Sketch what you consider to be a reasonable graph of the weights of the students before and after the program.

Teaching Matters: The level 4 activity introduces the notion of mapping the first value of a data pair to the second, with the equation giving the rule that explains how the first value is transformed into the second value. Exercises 23–26 require students to trace the mapping on the coordinate system and use the graph of the line to determine the missing x- or y-value. Exercises 27–30 require finding the same information from the equation that relates the data pairs. These important connections among the equation of a line, its graph, and how each states or portrays the relationship between the coordinate values of points on the line are fundamental, intuitive notions of a function.

Students at each level of these activities should explore in class discussions some nonlinear data that fit the relationship $y = ax^2$, $a > 0$, to make them aware that other relationships exist. By the end of the lesson, students should revisit the initial question concerning height and waist measurements as a class activity to apply the concepts under study.

Find an equation that relates the following data pairs. Use it to find the missing values.

27.

x	y
35	?
25	5
20	4
?	2
5	1

$[y = \frac{1}{5}x]$

28.

P	V
4	8
?	7
2.5	6.5
1.5	?
1	5

$[V = P + 4]$

29.

x	1.5	3	4	5.5	?	9
y	4	?	6.5	8	10	11.5

$[y = x + 2.5]$

30.

t	−12	−6	−2	0	?	8
Q	3	?	0.5	0	−1	−2

$[Q = -\frac{1}{4}t]$

WRITING AND EVALUATING VARIABLE EXPRESSIONS

Location in sample syllabus: Year 1, Unit 4

Major standards addressed: Algebra, connections

Objectives: ◆ To write and interpret an algebraic expression in terms of a given situation

◆ To develop an intuitive notion for the nature of a variable and the relationships among variable expressions

Prerequisite: Concept of perimeter

Sequence A will address level 1, and possibly level 2, of the lesson activities. Sequence B might begin with an activity similar to that in level 2 and then proceed to level 3 explorations.

Motivating question: A carpenter making three-legged and four-legged stools in a furniture factory was required to make approximately equal numbers of the two stools. On a day when materials were in short supply, she had eighty legs to use for making either type of stool. Since she had to set up her jigsaw differently for the type of stool made, she had to determine how many stools of each type to make before starting to work. How many of each type might she have made?

Level 1

Materials: pipe cleaners, tongue depressors, unit square tiles

Directions: Using pipe cleaners of length a, tongue depressors of length b, and tiles of unit length, form the following figures and state the lengths of all sides.

1. a triangle of perimeter $6a$

2. a rectangle of perimeter $2a + 6$

3. a regular hexagon of perimeter 12

4. a square of perimeter $8 + 4b$

5. a triangle of perimeter $a + 2b + 1$

6. a pentagon of perimeter $4b + 11$ with no two sides of equal length

   [Answers will vary.]

*Assessment Matters: **When the instructional emphasis is on concept building through situations reflecting real-world questions and activities, the assessment should be of a similar nature. Open-ended, holistically scored questions, interviews, observation of group work, testing with the use of physical models like those used in instruction, and student self-assessment are appropriate approaches.***

Write an expression for the perimeters of these figures:

7.   8.   9.

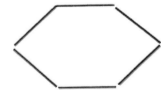

[$2a + b + 5$]   [$3(b + 1)$]   [$6a$]

Teaching Matters: In this activity, variables represent specific, but unknown, lengths. The meaning of the algebraic symbols and expressions arises through applying the concept of perimeter with concrete materials (e.g., 6a means a + a + a + a + a + a rather than 6 + a). Students intuitively practice combining like terms and learn informally that different-appearing expressions may be equivalent (e.g., different expressions for the perimeters will arise in exercises 8, 11, and 12). The physical constraints of the figures force students to associate the terms of some expressions differently from what they first suppose. (In exercise 5, a triangle cannot be formed with lengths a, 2b, 1 but can be formed with lengths a + 1, b, b or a, b, b + 1). To extend this activity, ask students to evaluate perimeter expressions for specific values of variables.

Teaching Matters: In this activity, variables have unknown values from a specific range of whole numbers. Students intuitively determine the domains of f and s. They are able to answer questions concerning the scores of the players by using the relationships among those scores, expressed as algebraic quantities. The idea that algebra deals with the relationships among variable quantities— independent of the specific values those variables might take—is a very powerful notion embedded in this activity and one that students will take a long time to appreciate fully.

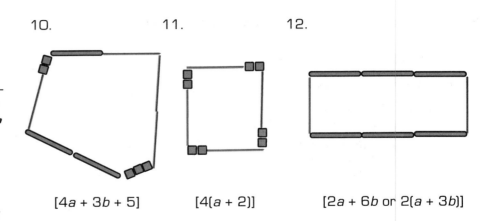

10.   11.   12.

[4a + 3b + 5]   [4(a + 2)]   [2a + 6b or 2(a + 3b)]

Level 2

Four players threw a pair of dice twice. On the first round of throws, Jeff got a score of f. Martha scored 4 more than Jeff. Quincy scored 2 less than Jeff, and Kwan scored 1 less than Martha. Use the diagram below to summarize the information for the first round to help you to answer the following questions.

	First round	Second round	Total for both rounds
Jeff			
Martha			
Quincy			
Kwan			

13. List the possible scores for Jeff.   [4, 5, 6, 7, 8]

14. List the possible scores for Quincy.   [2, 3, 4, 5, 6]

15. How many more points did Martha score than Quincy?   [6]

16. If Jeff and Kwan were teammates against Martha and Quincy, which team scored the most points on the first round?   [Jeff and Kwan]

On the second round, Jeff got a score of s. Martha scored 3 less than Jeff. Quincy scored 2 more than Jeff. Kwan scored 2 less than Quincy. Use this additional information to complete the diagram above, and then answer these questions:

17. Which players received the same score on the second round? [Jeff and Kwan]

18. List the possible scores that Jeff could have scored on the second round.   [5, 6, 7, 8, 9, 10]

19. Who scored the most total points for the two rounds?   [Kwan]

20. If Quincy scored 15 total points, list the total scores for the other players.   [Jeff, 15; Martha, 16; Kwan, 18]

Level 3

Prerequisite: Formula for the perimeter of rectangles and some familiarity with computers and BASIC

Equipment: A computer that uses BASIC commands

Complete this table for rectangles whose length is 3 more than their width. Write out the complete expressions as shown by the example.

Width (cm)	Length (cm)	Perimeter (cm)
4	4 + 3 = 7	2(4) + 2(4 + 3) = 22
6		
7.1		
9½		
x		

The following program prints out a table that lists the width, the length, and the perimeter of rectangles whose length is 3 more than their width. Complete the program by writing as line 40 the width, length, and perimeter expressions from the last row of the table above—preceded by the command PRINT and separated by commas. (Remember to separate the factors of a product with *.) Enter the program into a computer and run it to answer the following questions about rectangles whose length is 3 more than their width.

(When run, the output of the program below lists the dimensions of rectangles whose widths vary from 5 to 10 and increase in amounts of 0.5.)

```
10 REM RECTANGLES WHOSE LENGTH IS 3 MORE THAN THEIR WIDTH
20 PRINT "WIDTH", "LENGTH", "PERIMETER"
30 FOR W = 5 TO 10 STEP .5
40
50 NEXT W
```

21. What are the width and the length of a rectangle with a perimeter of 40?   [8.5, 11.5]

22. What is the perimeter of a rectangle whose width is 6.5?   [32]

23. What do you think the width and the length are of a rectangle with a perimeter of 43?   [9.25, 12.25]

24. How could you change the program above to generate a table with rectangles whose widths vary by 2s from 20 to 40?
[30 FOR W = 20 TO 40 STEP 2]

Level 4

Some Sequence B students will be able to do the following activity.

Activity directions:

Try to answer each set of questions first. Then consider the discussion questions about all three sets of exercises.

Set 1   Given: ❑ + ○ = 36

(❑ + ▲) + ○ = 55

25. What is the value of ○ + ❑?   [36]

26. What is the value of ❑ + (▲ + ○)?   [55]

27. What is the value of ▲?   [19]

28. What is the value of ❑?   [not enough information]

Teaching Matters: In the first set of information, we can answer questions about the symbol expressions by applying properties of numbers (which the symbols represent) and properties of equality. In the second set of information (unlike the first set), the real-world situation imposes restrictions on the domains of the variables. Though we cannot determine the values of x or y, the relationship among the variable expressions does allow us to conclude that Crewcut is one year older than the sum of the other two children. In the third set of information, the values of the letters are specified; they do not represent variables at all. However, the formula for the area of a triangle represents a generalization expressed in variables.

Set 2 In the same family, there are three children. Their ages in years are Shandra $x - 2$, Ned $y + 1$, and Crewcut $x + y$. Answer the following questions if you can. Use a spreadsheet to help you decide.

29. How old is Shandra?   [not enough information]

30. How old is Ned?   [not enough information]

31. What do you know about x and y?   [$x > 2$, $y > -1$]

32. Who is the oldest of the three?   [Crewcut]

Set 3 Use the formula $A = \frac{1}{2}bh$ to find the areas of triangles with these dimensions:

32. $b = 2$ cm, $h = 4.5$ cm   [$A = 4.5$ cm²]

33. $b = 9$ ft., $h = 1.2$ ft.   [$A = 5.4$ ft.²]

Discussion questions:

34. Letters and symbols are used to represent unknown quantities in the question sets above. What is different and what is the same about the way in which these letters and symbols are used?

35. Some of the questions above can be answered and some cannot. What kind of information did you use to answer the questions that were answerable?

VARIATION

Location in sample syllabus: Year 1, Unit 11

Major standards addressed: Algebra, functions

Objective: To interpret direct and inverse variation in terms of rates and constants of variation

Prerequisites: Knowledge of linear relationships of the form $y = ax$

Sequence A will address level 1, and possibly level 2, of the lesson activities. Sequence B might begin with data from activities similar to level 1 to introduce the notions of variation and then move on to interpretive situations like those in level 2 before beginning level 3 activities.

Motivating question: The following graphs represent the progress of a 100-meter race run by a mother against her daughter (assuming that each ran at a constant rate). The mother agreed to race only if she were given a 20-meter head start. The graphs give the distance from the starting line of each contestant t seconds after the start. Answer the following questions from the information given by the graph.

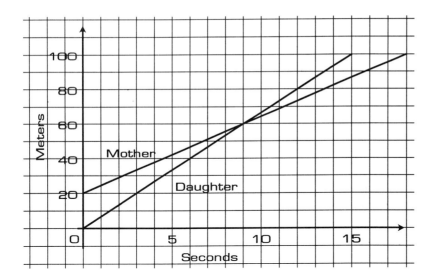

a. How far from the starting line was each runner after 2 seconds? [~29 m, ~7 m]

b. At what time did each runner reach the 40 m mark? [~4.5 s, 6 s]

c. What does the point of intersection of the two graphs mean? When did it occur? At what location?
[The racers were side-by-side; 9 s; 60 m from the starting line]

d. Who won the race? By how much? [The daughter; ~13 m or 3 s]

e. At what time did each person cross the 100 m line? [15 s, 18 s]

f. What was each runner's speed?
[Mother: 4.5 m/s; daughter: 6.67 m/s]

g. Write an equation for the daughter's graph. [$d = 6.67t$]

h. Write an equation for the mother's graph. [$d = 4.5t + 20$]

i. Which graph represents a direct variation? [The daughter's graph]

Level 1

Materials: Springs, spring balances, gram weights, number balance (primary tool) or simple lever, calculators, centimeter graph paper

Activity A

Students discover Hooke's law by suspending weights from a spring and noting the displacement (stretch) of the spring. By recording the weight in grams versus the displacement in centimeters, students will collect data pairs (w, d) that they can then plot to determine that the relationship is linear. Next, by finding the ratios d/w, they can determine the constant of variation for the spring and the graph, establishing the equation of the linear graph and the constant quotient property of direct variations.

w (g)	0	50	100	200	250	325
d (cm)	0	1.2	2.5	4.9	6.0	7.7

d/w: —, 0.024, 0.025, 0.0245, 0.024, 0.024
direct variation: $y = kx$
k is the constant of variation
constant quotients: $y/x = k$

Activity B

The number balance contains pegs equally spaced from the fulcrum and from which weights can be suspended. Have students suspend a 200 g weight from a peg 5 units from the fulcrum. Next, have students use a spring balance to exert a force from different pegs on the other side of the fulcrum sufficient to balance the 200 g weight on its 5-unit moment arm. Students will record data pairs (f, s) representing the force f in grams and the distance s from the fulcrum necessary to achieve a balance (raise the weight on the other end of the lever).

By plotting data pairs, students will discover the shape of an inverse variation relationship (as the distance s increases, the amount of force f decreases). Next, have students find the product of each data pair to establish the constant product property of inverse variations and the constant of inverse variation.

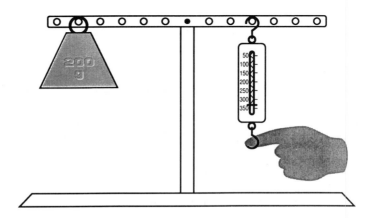

inverse variation: $y = k/x$
k is the constant of variation
constant products: $xy = k$

Teaching Matters: The emphasis in level 1 is to establish the relationships through direct experimentation, relating the graphs to activities in which students experience how one quantity varies with respect to another. The constants of variation have meaning in terms of the experimental conditions. Problems that students are later asked to solve should be phrased in contexts that relate the problems to similar experimental situations or interpretations. Extensions of these activities could ask students to determine the effects on the outcomes (and graphs) of changing the springs (activity A) or changing the weight or location of the stationary weight (activity B).

Using the constant quotient and constant product properties, students can establish methods consistent with the following formulas for use when a variation is known and one data pair is given:

$$k = \frac{y_1}{x_1} = \frac{y_2}{x_2} \qquad\qquad k = x_1 y_1 = x_2 y_2$$

Level 2

After completing level 1, some students will complete the following activity.

Directions: Each of the following situations involves a rate that relates two quantities that vary directly with one another.

a. For each situation, label a horizontal and a vertical axis as directed, and then use the rate to find and plot five ordered pairs that satisfy each variation.

Water flowing into a bucket at 2 liters/min

horizontal axis —time (min)
vertical axis—volume (liters)

Candy selling at $5 per box

horizontal axis—boxes of candy
vertical axis—revenue (dollars)

30 percent off sale at the clothing store

horizontal axis—original price   vertical axis—sale price

b. Write an equation in words that shows how the quantity on the vertical axis is related to the quantity on the horizontal axis.

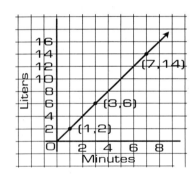

$$V = 2 \times t$$
volume = 2 × time

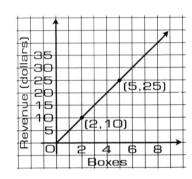

$$R = 5 \times n$$
revenue = 5 × no. of boxes

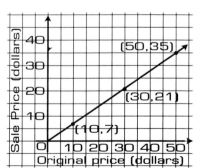

$$S = 0.7 \times P$$
sale price = 0.7 × original price

Students should now be able to answer the questions about the graphs introduced by the *motivating question.*

Teaching Matters: **When the topic of inverse variation is revisited in a subsequent course, students would use a graphing utility to compare data points to the graph of y = k/x, for the value of k they have found. It would also be the time to begin to think about errors. Where is the error the greatest? Compare the absolute error to the relative error.**

Teaching Matters: **Students are introduced to the notion of direct variation applications involving rates. Using the given rate, students plot points that represent conditions relating the two variable quantities. By interpreting a linear graph, students are introduced intuitively to the notion of slope as rise over run and the notion of rate as the constant of variation in an equation relating the two variable quantities.**

Teaching Matters: **After students discover the relationships on the graphs, they should be directed to verify that each equation is satisfied by the coordinates of the points originally plotted.**

Teaching Matters: **Through small-group work on similar problems, students will extend their ideas about direct variation relationships while gaining additional insights into related graphical concepts. This sets the stage for formal development of these ideas in this and subsequent courses (e.g., slope in the next unit).**

Level 3

Students graphically explore rate problems by extending the notion of direct variation as a constant quotient, or *rate*, when applied to situations involving more than one rate.

Example problem:

Akimura can paint a large room in 2 hours. Benito can paint a similar large room in 3 hours. How long would it take them working together to paint the interior of a house containing five large rooms?

Through a guided discussion, students are led to the realization that the number of rooms painted by each person varies directly according to time. By considering how many rooms Akimura can paint in 0, ½, 4, or 8 hours, students can see that a linear graph represents this variation. A similar graph can be constructed for Benito's painting.

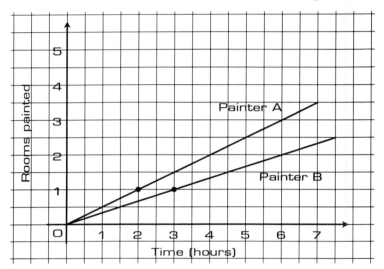

Next, raise the question about finding the combined rate of their working together. By considering how much each does in one hour, students will suggest intuitively the addition of ordinate values to give a data point on the graph of their combined rate. Knowing that the origin must also satisfy this variation, they can draw the linear graph. Interpreting the appropriate point on this combined graph answers the question of the problem.

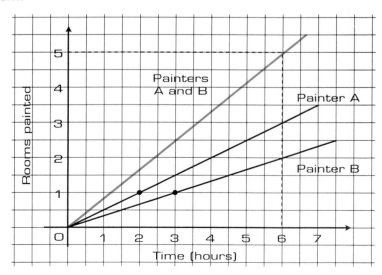

Level 4

Some Sequence B students will be able to progress to the following activity.

Students should be able to solve problems involving rates both algebraically and graphically. The strength of using these two approaches to the same type of problem is that they are mutually supportive. For example, the work accomplished by each painter in the preceding problem is a term in the following equation, corresponding to the ordinate values of the graphs above (as the number of rooms painted by each painter in t hours).

Let t represent the time in hours required for Painter A and Painter B to paint five rooms together.

Rate of Painter A = ½ room per hour
Rate of Painter B = ⅓ room per hour

Word equations:

$$\text{rooms painted by A} + \text{rooms painted by B} = 5 \text{ rooms}$$
$$\text{rate of A per hour} \times t \text{ hours} + \text{rate of B per hour} \times t \text{ hours} = 5 \text{ rooms}$$
$$\tfrac{1}{2} \cdot t + \tfrac{1}{3} \cdot t = 5$$
$$t = 6 \text{ hours}$$

As an extension of direct variation relationships for linear functions, consider how one variable might vary as the square or square root of another variable. Have students explore how the surface area, A, of a cube varies according to its edge length, s.

a. Have students first determine a table of values for (s, A) and graph them to verify that A does not vary directly as s.

b. Next, rewrite the equation for the surface area, $A = 6s^2$, as the constant quotient $A/s^2 = 6$ and ask students how they might interpret the constant quotient as a direct variation.

c. Have students then complete a table of values for (s^2, A) and graph them. Students should verify that each data point satisfies the constant quotient $A/s^2 = 6$ and that the graph is linear.

d. Then develop the equation $A_1/s_1{}^2 = A_2/s_2{}^2$ and investigate how a proportion can be used to answer questions about relationships involving this type of variation.

e. Extend this more general idea to other functions, such as to the square root function.

Teaching Matters: The following experiment illustrates direct variation with squared functions. Collect a number of jar lids of different kinds. Have students measure the diameter and approximate the radius from that measure. Now fill the lid one unit deep with BBs or similar material. Use the number of BBs as a measure of area A and compute A/r^2 for each lid. Construct a graph of the ordered pairs (r, A) and (r^2, A). Ask students what measurement they need so that the slope of the line associating the pairs (r^2, A) is approximately π. [Students need to know the diameter of a BB to determine its radius and "area."]

INDUCTIVE AND DEDUCTIVE REASONING WITH EXTERIOR ANGLES

Location in sample syllabus: Year 2, Unit 2

Major standards addressed: Reasoning, synthetic geometry

Objective: To distinguish between inductive and deductive reasoning and use each to demonstrate generalizations about exterior angles

Prerequisites: ♦ The degree sum of the interior angles of a convex polygon with n sides is $(n-2) \times 180$.
♦ The degree sum of the angles of a linear pair is 180.

Materials: Protractors, straightedges

Sequence A will address level 1, and possibly level 2, of the lesson activities. Sequence B will address levels 1–4.

Motivating question: An exterior angle of a polygon is formed by extending a side at a vertex, as shown at the right. We can find the sum of the exterior angles, one at each vertex, for a polygon. In previous studies, we have seen that the degree sum of the interior angles of a polygon of n sides is $(n-2) \times 180$. What do you suppose happens to the sum of the exterior angles of a polygon of n sides as n increases? How can you be sure?

Teaching Matters: This should be an open exploration. Students could use Logo or a Supposer-type software to make conjectures about exterior angles for different figures. A simple experiment with a pencil is to point the pencil along one ray, slide it up until the eraser is at the vertex, rotate it through the exterior angle, and slide it to the next vertex, repeating until the pencil is returned to the original ray. The pencil will make one full rotation, suggesting that the degree sum of the exterior angles is 360. Discuss with students how they can be certain that this is always the case. Point out that checking many polygons is an example of inductive reasoning. Be sure they recognize that a drawing or measuring utility may help in formulating and verifying conjectures, but it cannot provide a proof.

Directions for the initial activity:

Study the polygon shown and make at least three observations about the exterior angles. [Each is a supplement of an interior angle; in general, exterior angles may be obtuse or acute; there are $n = 6$ of them; they make a pinwheel design; by extending rays for both sides at a vertex, we could form twice as many angles.]

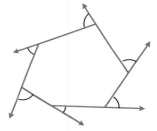

exterior angles

Level 1

1. Begin with a simple case of one exterior angle of a triangle. Draw a triangle and form one exterior angle as shown. Find the degree measure of the four angles shown and record your data in a chart.

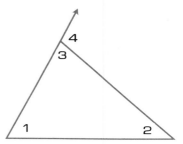

∠1	∠2	∠3	exterior ∠4	

Look for patterns in the chart.
[$\angle 1 + \angle 2 + \angle 3 = 180$; $\angle 3 + \angle 4 = 180$]

2. Find the sum of $\angle 1$ and $\angle 2$. Put this number in the last column. What do you notice? [$\angle 1 + \angle 2 = \angle 4$]

3. When we look at many cases and see a common pattern or generalization, we are using **inductive reasoning.** What generalization does inductive reasoning suggest from your results in exercise 2? [The measure of an exterior angle equals the sum of the measures of the two nonadjacent angles.]

4. Test your conjecture with more triangles, looking for an exception or **counterexample.**

5. The data should suggest that our generalization is always true. But we can never test all possible triangles. We need more convincing reasons to believe that the pattern is always true. Consider the following argument. Read it, discuss it with your partner, then try to reconstruct it yourself.

   We know for any triangle that $\angle 1 + \angle 2 + \angle 3 = 180$. Why?

   Also, we know that an exterior angle forms a linear pair with its adjacent interior angle.

   $$\angle 4 + \angle 3 = 180$$

   Examining the two equations, we see that the angles that are added to $\angle 3$ in each equation must be equal amounts.

   $$\angle 1 + \angle 2 = \angle 4$$

6. The reasoning above demonstrates that for any triangle, the measure of an exterior angle is the sum of the measures of the two nonadjacent interior angles. It is a general argument based on logic and principles that we have already accepted as true. This type of argument is called **deductive reasoning.**

Follow-up activities:

In exercises 7 and 8, identify whether inductive or deductive reasoning is used.

7.
$$1 = 1$$
$$1 + 3 = 4$$
$$1 + 3 + 5 = 9$$
$$1 + 3 + 5 + 7 = 16$$
$$1 + 3 + 5 + 7 + 9 = 25$$
$$1 + 3 + 5 + 7 + 9 + 11 = 36$$

   Gina concluded that the sum of the first n consecutive odd numbers is n^2.

8. Colin used the following argument to reason why the sum of two even numbers is even:

   An even number has a factor of 2, so it can be written $2k$, where k is some integer. The sum of two even numbers could be represented as $2k + 2n$. By the distributive property, $2k + 2n = 2(k + n)$. Since k and n are integers, then so is $k + n$. Therefore, $2(k + n)$ is another even number, so the sum is always even.

9. Describe and then carry out an experiment that uses inductive reasoning to find the sum of the exterior angles of a triangle. State the generalization. [Measure and total the exterior angle measures for many triangles. The sum of the degree measures of the exterior angles of a triangle is 360.]

10. The following deductive argument demonstrates the sum of the measures of the exterior angles of any triangle. It uses the *triangle exterior angle property* (established earlier)—each exterior angle of a

*Teaching Matters: **One** means to integrate communication into mathematics and to assist the development of reasoning skills is to have students read a deductive argument (proof), explain it to another person, and reconstruct it (possibly with clues) at another time. Try this strategy with exercise 5. Class discussion should assist students to distinguish between inductive and deductive reasoning and to understand the limitations of inductive reasoning.*

triangle equals the sum of the two nonadjacent interior angles. Fill in the missing statements below.

In the diagram,

$\angle a = \angle e + \angle f$ by the $\triangle$ exterior angle property

$\angle b = \angle d + \angle f$

$\angle c = \angle d + \angle e$

$\angle a + \angle b + \angle c = \angle e + \angle f + \angle d + \angle f + \angle d + \angle e$ by adding equals from above

$\angle a + \angle b + \angle c = 2\angle d + 2\angle e + 2\angle f$ by _____

$\angle a + \angle b + \angle c = 2(\angle d + \angle e + \angle f)$ by _____

$\angle a + \angle b + \angle c = 2(180)$ by _____

$\angle a + \angle b + \angle c = 360$ by multiplication

Level 2

After completing level 1 activities, some students will complete the following activity.

11. Describe and carry out an experiment using inductive reasoning to find the sum of the exterior angles of a pentagon. State the generalization.

12. The following deductive argument demonstrates the sum of the measures of the exterior angles of any pentagon. It uses the *linear pair property*—angles forming a linear pair total 180—and the *polygon sum property*—the sum of the degree measures of the interior angles of a convex n-gon is $(n - 2) \times 180$. Fill in the missing statements below.

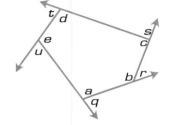

$\angle a + \angle b + \angle c + \angle d + \angle e = 3(180)$ by _____

$\angle a + \angle q = 180$ by the linear pair property

$\angle b + \angle r = 180$

$\angle c + \_\_\_ = 180$

$\angle d + \_\_\_ = 180$

$\_\_\_ + \_\_\_ = 180$

$(\angle a + \angle b + \angle c + \angle d + \angle e) + (\angle q + \angle r + \angle s + \angle t + \angle u) = 5(180)$

by addition of equals

$3(180) + (\angle q + \angle r + \angle s + \angle t + \angle u) = 5(180)$

by substitution

$\angle q + \angle r + \angle s + \angle t + \angle u = 2(180)$

by _____

13. Consider the pentagon drawn at the right. Its exterior angles measure 90, 90, 135, 90, and 135 degrees.

    a. Compare the sum of these exterior angles with results from exercises 11 and 12.

    b. What do you conclude? [This is a counterexample to our argument. We have to restrict our argument to convex pentagons.]

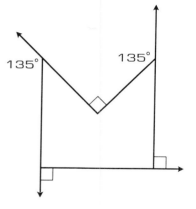

14. a. What do you expect the sum of the exterior angles of a convex hexagon to be?

    b. Extend your argument from exercise 12 to apply to any convex hexagon.

    c. What do you expect the sum of the exterior angles of a convex n-gon to be? Explain.

Level 3

15. Draw several convex polygons. For each, find the sum of the exterior angles. Compare data around the class. What generalization can you make? What sort of reasoning are you using?

16. Write a deductive argument to prove that the sum of the degree measures of the exterior angles of any convex n-gon is 360. Work with a partner.

17. Suppose we define the total angle exterior to a polygon at each vertex as the sum of the measures of the angles formed by extending the sides. What, if anything, can you conclude about the sum of the measures of these angles?

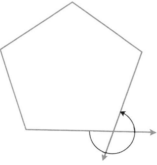

Form a group of three students and explore various situations. If you can state a generalization, try to create an argument that proves it. Compare your expression for the sum of the measures of the total exterior angles of an n-gon to those of other groups; compare your method for establishing it. [The sum of the degree measures of the total angles exterior at each vertex of an n-gon is $180n + 360$.]

Level 4

This level assumes that students have completed levels 2 and 3 and have had some experience with investigating concave polygons that do not apparently fit the pattern.

18. Consider concave pentagon *ABCDE* again. Beginning at *A*, headed south, imagine walking around the polygon. Note that each turn corresponds to an exterior angle of the polygon. How many left turns did you make? How many right turns did you make?

19. If we consider a right turn to be the opposite of a left turn, then we might assign to that angle a negative value. Do this in the exercise above and determine the sum of the exterior angles.

20. Try the same experiment with other concave polygons. State a generalization. What kind of reasoning have you just used?

Teaching Matters: **The activities in level 3 of this sequence depend more heavily than Sequence A activities on abstract algebraic arguments. The power of these activities, particularly exercise 17, is that they return students to the major generalizations of this lesson and to the processes—inductive and deductive reasoning—that established those generalizations. A class will typically come up with at least three ways of establishing the sum in exercise 17, which result in different algebraic forms of the generalization.**

Teaching Matters: **This activity should spur an open discussion with the focus on the anomaly at D. Students may also draw, consider, and measure other concave polygons to see if there is some other regularity.**

Location in sample syllabus: Year 2, Unit 6

Major standards addressed: Synthetic geometry

Objective: To understand and apply the ratio property for three-dimensional figures (If two figures are similar with a ratio of similarity of k, then corresponding lengths have ratio k, corresponding areas have ratio k^2, and corresponding volumes have ratio k^3.)

Prerequisites:
- ◆ Familiarity with similar figures on the plane and in space
- ◆ Understanding distinctions of one-, two-, and three-dimensional measures
- ◆ Knowledge of ratio of similarity and correspondence in similar figures
- ◆ Knowledge of surface area and volume formulas for prisms, cylinders, pyramids, cones, spheres

Sequence A will address level 1, and possibly level 2, of the following lesson activities. Sequence B will focus on levels 3 and 4.

Teaching Matters: **Guide students' initial discussion with questions such as these: What is the relationship between the model and the plane? [Same shape; all lengths proportionally larger; similar] What ratio relates the two figures? [1:60] Do you expect that 40 × 60 = 2400 (mL) of paint would be enough? Why not? What simpler figures could we consider to understand the relationships involved in the questions above? [Cubes, prisms, pyramids, spheres]**

Motivating question: Denver Aviation High School is going to receive a surplus aircraft for display on the lawn. The student booster club has received a model of the airplane, which is 15 cm long. The actual airplane is about 9 m long. Knowing that the plane will need painting, the booster club paints the model in the colors to be used and observes that 40 mL of paint is needed. How much paint will be required for the plane? They also speculate about the volume of the plane compared to the volume of the model. How did they answer these questions?

Directions for the initial activity:

1. In each of the following situations, identify the use of one-, two-, or three-dimensional measures.

   a. A box is filled with goods, covered with wrapping paper, and tied with string.　[3-D, 2-D, 1-D]

   b. A pillow is filled with stuffing, covered with a case, and decorated with lace around the edges.　[3-D, 2-D, 1-D]

   c. The wings of an airplane have span (distance from tip to tip), require paint to cover, and contain gas tanks.　[1-D, 2-D, 3-D]

2. Elementary classrooms sometimes use "big books," magnified versions of child-sized story books, for group story times. Consider the following changes. Which appear to be the best magnifications to produce a "big book"?　[e]

   a. twice as tall

   b. twice as wide

   c. twice as thick

   d. twice as thick and twice as tall

   e. twice as tall, twice as thick, and twice as wide

   f. twice as thick and twice as wide

Level 1 (teacher-directed)

3. Give students a net of a cube that measures 4 cm along each edge. Ask them to create a net of another cube in which each edge is magnified by a factor of 3. Have them cut out each net and fold them into cubes.

4. Ask students to find and record the total surface area and volume of each cube.
   [Original: 96 cm^2, 64 cm^3; magnified: 864 cm^2, 1728 cm^3]

5. Ask students to use their results from exercise 4 to find each of the following ratios, expressed in simplified form.   [3:1, 9:1, 27:1]

   a. $\dfrac{\text{edge of enlarged cube}}{\text{edge of original cube}}$

   b. $\dfrac{\text{surface area of enlarged cube}}{\text{surface area of original cube}}$

   c. $\dfrac{\text{volume of enlarged cube}}{\text{volume of original cube}}$

Level 2

6. A cube has edges measuring s cm. The cube is enlarged by a factor of k. Write these expressions:

   a. the length of an edge of the enlarged cube   [sk]

   b. the surface areas of the original and the enlarged cubes
      [Original: $6s$; enlarged: $6k^2s$]

   c. the volumes of the original and enlarged cubes
      [Original: s^3; enlarged: k^3s^3]

   d. the simplified ratios below:

      (1) $\dfrac{\text{edge of enlarged cube}}{\text{edge of original cube}}$

      (2) $\dfrac{\text{surface area of enlarged cube}}{\text{surface area of original cube}}$

      (3) $\dfrac{\text{volume of enlarged cube}}{\text{volume of original cube}}$ [k, k^2, k^3]

7. A rectangular prism has dimensions $a \times b \times c$. It is enlarged by a factor of k. Write these expressions:

   a. the lengths of the edges of the enlarged prisms   [ka, kb, kc]

   b. the surface areas of the original and the enlarged prisms
      [Original: $2ab + 2bc + 2ca$; enlarged: $k^2(2ab + 2bc + 2ca)$]

   c. the volumes of the original and the enlarged prisms   [$abc; k^3abc$]

   d. the simplified ratios below   [k, k^2, k^3]

      (1) $\dfrac{\text{edge of enlarged prism}}{\text{edge of original prism}}$

      (2) $\dfrac{\text{surface area of enlarged prism}}{\text{surface area of original prism}}$

      (3) $\dfrac{\text{volume of enlarged prism}}{\text{volume of original prism}}$

8. Complete: When two prisms are similar, with a ratio of proportionality k, their surface areas have a ratio of proportionality _____ and their volumes have a ratio of proportionality _____.   [k^2, k^3]

*Teaching Matters: **Exercise 1** helps students recall distinctions among 1-, 2-, and 3-dimensional figures. Exercise 2 helps students recognize that 3-dimensional figures are similar only when the larger is a magnified version of the smaller in all three dimensions. Other examples of models of life-sized objects are model cars, houses, and trains. Objects created in different-sized versions are nesting toys, nuts and bolts, science beakers, and stack tables.*

*Teaching Matters: **Students** should explain their ratios in exercise 5 in terms of the magnification factor and the number of dimensions involved in the type of measure. You may wish to have students investigate these ideas through nets and models of figures (e.g., pyramids and cylinders) and other magnification or reduction factors. Students should eventually discover through class discussion the ratio property for 3-dimensional figures.*

*Teaching Matters: **In both** levels 1 and 2, students should be asked to consider similar cases with $0 < k < 1$ (reductions). They should observe that the general relationships between dimensions are not changed. Students may want to investigate other figures to be certain of the generality of the ratio property. Select activities that are appropriate to the level of abstraction (concrete, numerical, formula) at which students are working. Students should now answer the motivating question of the lesson.*

♦　　♦　　♦　　♦　　♦　　♦　　♦　　♦

Teaching Matters: The ratio property for 3-dimensional similar figures has many applications in science and literature. For example, the Lilliputians in Jonathan Swift's Gulliver's Travels *figured that since Gulliver was twelve times taller than they were, he would need 12³, or 1728, times the amount of food. In people, this property explains why giants do not exist and why very tall people have skeletal problems. (The weight of a person is proportional to his or her volume or height cubed. That weight is supported on their bones whose cross-sectional area is proportional to height squared.)*

A related phenomenon is the "shivering child syndrome." Small children can be observed shivering at a swimming pool on days when adults are perfectly comfortable. (Heat is generated according to the mass or volume of the individual, whereas heat is lost according to a person's surface area. As a person's size increases, the ratio volume:surface area increases, meaning that overheating becomes a problem with increasing size.) Have a group of students read D'arcy Wentworth Thompson's "On Magnitude," in The World of Mathematics, *or J. B. S. Haldane's "On Being the Right Size," and describe other instances in nature of how the size of animals affect what they do well or poorly.*

9. Recall that the formula for the volume of a sphere is $V = \frac{4}{3}\pi r^3$ and the formula for its surface area is $S = 4\pi r^2$. Suppose a sphere of radius x is enlarged by a factor of k:

   a. Find the surface areas and the volumes of the original and the enlarged spheres.　[Original: $4\pi x^2$, $\frac{4}{3}\pi x^3$; enlargement: $4\pi k^2 x^2$, $\frac{4}{3}\pi k^3 x^3$]

   b. Can the generalization in exercise 8 be extended to spheres? If so, how? If not, why not?　[Yes; replace "prism" with "sphere."]

Level 3

10. A rectangular prism of dimensions $4 \times 6 \times 9$ cm is compared with another similar prism in which all dimensions are enlarged by a factor of k. Write these expressions:

    a. the length of the edges of the second prism　$[4k, 6k, 9k]$

    b. the surface areas of the original and the second prisms $[228 \text{ cm}^2; 228k^2 \text{ cm}^2]$

    c. the volumes of the original and second prisms $[216 \text{ cm}^3; 216k^3 \text{ cm}^3]$

    d. the simplified ratios below:

    (1) $\dfrac{edge\ of\ second\ prism}{edge\ of\ original\ prism}$　　(2) $\dfrac{surface\ area\ of\ second\ prism}{surface\ area\ of\ original\ prism}$

    (3) $\dfrac{volume\ of\ second\ prism}{volume\ of\ original\ prism}$　　$[k, k^2, k^3]$

11. Repeat exercise 10 for a cone with radius 3 cm and height 8 cm, reduced by a factor of k to a similar cone. Write these expressions:

    a. the radius and the height of the second cone　$[3k, 8k]$

    b. the surface areas of the original and second cones $[\ 80.53 \text{ cm}^2, (80.53 \text{ cm}^2)/k^2\ ]$

    c. the volumes of the original and the second cones $[75.40 \text{ cm}^3, (75.40 \text{ cm}^3)/k^3\ ]$

    d. the simplified ratios below:　　　　$[1/k, 1/k, 1/k^2, 1/k^3]$

    (1) $\dfrac{radius\ of\ second\ cone}{radius\ of\ original\ cone}$　　(2) $\dfrac{height\ of\ second\ cone}{height\ of\ original\ cone}$

    (3) $\dfrac{surface\ area\ of\ second\ cone}{surface\ area\ of\ original\ cone}$　　(4) $\dfrac{volume\ of\ second\ cone}{volume\ of\ original\ cone}$

Divide the class into working groups of three or four students to explore whether the results in exercises 10 and 11 can be generalized for the following figures for which formulas are known: *cube, prism, cylinder, pyramid, cone, sphere.* Assign each group two different figures to explore according to the following directions (results are to be compiled in a table on the chalkboard or the overhead projector).

12. For each figure—

    a. choose variables (different from those used in the common formulas) for the dimensions of the original figure;

    b. imagine an enlargement or a reduction of magnitude k in every dimension of the figure and write expressions for the dimensions of the enlarged (reduced) image of the figure;

    c. write expressions for surface areas and volumes for the original figure and its image;

d. find the following ratios in reduced form:

(1) $\dfrac{lengths\ for\ image}{lengths\ for\ original\ figure}$ (2) $\dfrac{surface\ area\ of\ image}{surface\ area\ of\ original\ figure}$

(3) $\dfrac{volume\ of\ image}{volume\ of\ original\ figure}$

Level 4

Some Sequence B students will be able to do the following activities.

Exploration question: How do the volumes and the areas of geometric figures vary when they are rescaled by different proportions in different dimensions?

13. Imagine a box of height 10 inches, width 9 inches, and length 12 inches. It is transformed into another box by taking half the height, tripling the width, and doubling the length.

    a. Find the dimensions of the new box.
       [$h = 5$ in., $w = 27$ in., $l = 24$ in.]

    b. Find the volumes of the original and the new boxes.
       [1080 in.^{3}, 3240 in.^{3}]

    c. Find the ratio of the box volumes (new:original).   [3:1]

    d. How does the ratio of the volumes compare to the factors in the transformation?
       [The ratio of the volumes is the product of the size changes: $0.5 \times 3 \times 2 = 3$]

    e. Repeat the steps above for a box with dimensions a, b, c. Assume that the dimensions are rescaled by factors k, m, n, respectively.

14. The area of a circle is given by $A = \pi r^2$. We can consider each factor of r as representing a measure in one of two perpendicular dimensions, just as l and w are measures in perpendicular dimensions for a rectangle.

    a. We can think of an ellipse as a circle that is rescaled horizontally by one factor and vertically by another. For simplicity, imagine starting with a unit circle ($r = 1$). Transform it by a (horizontally) and b (vertically). What do you conjecture the area formula for an ellipse to be?   [$A = \pi ab$]

    b. Check your conjecture graphically by drawing a unit circle on graph paper.

       (1) Let $a = 3$ and $b = 5$. Draw an ellipse with these dimensions on graph paper and determine its area by counting squares (estimating fractions for incomplete squares).

       (2) Compare your estimated area in (1) with your conjecture in part a.

Teaching Matters: In comparing the summary of class results from the table, students should discover the ratio property for similar figures and be able to state that generalization in written form.

Teaching Matters: For enrichment, have students investigate the problem presented in this Greek myth: A plague terrorized the ancient city of Délphos. The citizens consulted the oracle, who told them that the plague would be halted by doubling the size of the altar of Apollo. Assume the altar was a rectangular solid. The Delphians doubled the length, the width, and the height. The plague continued.

a. What error did the Delphians make?

b. By what factor should they have multiplied the dimensions of the altar? [$\sqrt[3]{2}$] Explain.

c. Suppose that the altar was not a rectangular solid. What should they have done to create a similar altar with double the volume?

POLYNOMIALS AND GEOMETRIC MODELS

Location in sample syllabus: Year 3, Unit 5

Major standards addressed: Algebra, geometry, problem solving, communication, connections

Objectives: ◆ To write a polynomial representing a geometric situation

◆ To find the product of one polynomial by another

◆ To explore a polynomial function representing a geometric situation

Materials: Algebra tiles or their equivalent

Prerequisites: Laws of exponents, max-min characteristics of the graph of a function

Sequence A will address level 1, and possibly level 2, of the following lesson activities. Sequence B will provide exploratory work similar to levels 1 and 2 and then will address levels 3 and 4.

Motivating question: A rain gutter is to be made from a rectangular piece of metal ten feet long and twelve inches wide by turning up equal edges of the width to form the base and sides of the gutter. What are the base and the height of such a gutter that can carry away the largest amount of rainwater?

Level 1 (teacher-directed)

Directions:

1. Give each student a plain sheet of paper and instruct them to fold the paper lengthwise to form the cross section of a rain gutter that can carry away the largest amount of water.

2. Have students compare their models of rain gutters and ask them how they can decide which models are better for carrying away rain.

3. After students have determined that the cross-sectional area of the gutter is the appropriate measure of its rainwater capacity, have each student measure the base and the height of their models and determine the appropriate area.

4. List the dimensions and the areas of the student gutter models with the relatively largest areas and pose the question about the best theoretical model that could be made.

5. With students' assistance, derive expressions for the base, the height, and the area of the gutter model, using x as the height. Indicate that the area is given by a function $A(x) = x(11 - 2x)$ whose right side represents a polynomial in factored form.

6. Decide with the students that the height x of the best theoretical gutter model can be found by determining the x-value that will make the polynomial $11x - 2x^2$ as large as possible.

7. Explore the gutter model through a table of values and through the graph of $A(x) = 11x - 2x^2$ to answer the gutter question relative to the paper models and to begin the discussion of polynomials and polynomial functions.

Follow-up activities:

8. Introduce students to algebra tiles, asking them to represent with the tiles different areas expressed as polynomials in both standard and factored forms:

$$x^2 + 3x \qquad 2(x + 1) \qquad (x + 2)(x + 4)$$

9. Give students algebra tiles arranged in rectangles and ask them to express the areas in two ways: (1) as the dimensions of rectangles and (2) as the areas of tiles of similar shapes.

$$(x + 3)(x + 2) = x^2 + 5x + 6$$

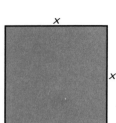

10. Develop a generalized rule for multiplying polynomials by illustrating the distributive property with algebra tiles. Then apply this property to the multiplication of polynomials without the use of tiles.

Level 2

Use the same motivating activity (rain gutter paper folding) to introduce polynomials. Then develop a generalized rule for multiplying polynomials by using an area model such as the following and expressing the area as $L \times W$ or as the sum of the individual areas.

	x	y	z
a	ax	ay	az
b	bx	by	bz

$$(x + y + z)(a + b) = ax + ay + az + bx + by + bz$$

Have students explore the area model for multiplying polynomials by completing exercises of the following types:

11. Fill in the missing parts of these diagrams. Then show the product and the expansion of the binomials represented by each figure.

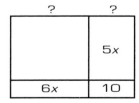

12. Ask students to multiply polynomials by the generalized rule without the aid of figures.

13. Have students work in groups to answer the rain gutter problem posed by the motivating question of this lesson, writing up a group analysis of the problem and the group answer.

Level 3 (student-centered)

14. Draw a figure that represents the area $(a + b)^2$. Then subdivide this figure into four smaller rectangles that correspond to the terms of the polynomial $(a + b)^2$ when expanded. Label the dimensions of each rectangle on the figure, and inside each rectangle list its area.

Teaching Matters: Class-room-ready activities that capitalize on an area model to enable all students to develop a conceptual understanding of linear equation solving and quadratic factoring can be found in Activities for Implementing New Curricular Themes from the Agenda for Action (Hirsch 1986).

15. a. Make paper models of each of the following figures:

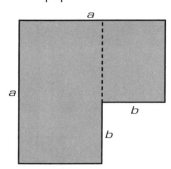 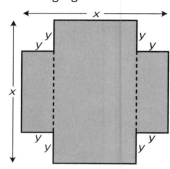

b. Write the area of each figure as a polynomial.

c. Cut each figure along the dashed line(s) and arrange the pieces into a new rectangle.

d. Write the area of this new rectangle as a product of its length and width.

e. Expand this product to show that the area of the newly formed rectangle is equal to the area of the original figure.

16. Write the area of this figure as the sum of the areas of a parallelogram and a triangle. Then expand and simplify this expression to show that it is equal to the area of the trapezoid.

17. Draw a diagram of a cube whose volume is given by the expression $(a + b)^3$. Draw a second diagram that shows how this cube can be sliced into eight rectangular prisms whose volumes correspond to the expansion of the expression $(a + b)^3$.

Level 4

After working through the rain gutter model and the activities of level 3, students will work in groups to solve the following problems, using a graphing utility. Each group is expected to submit one final written report of the problem as a detailed summary of its work.

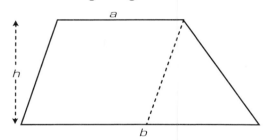

Teaching Matters: In level 3, students extend the area model for multiplying polynomials by introducing negative terms and reconstituting figures to correspond to the factored and expanded forms of polynomials. Students are also expected to work with less teacher direction. In level 4, students generate polynomials for examining max-min problems and investigate the behavior of these functions through graphing utilities. In examining the graphs of these polynomials for the conditions resulting in the maximums, students can be asked to find a graphical interpretation for the factored form of the polynomial. Relating how the size of the factors varies as the independent variable can be further illustrated with a spreadsheet.

18. There are 200 rods of fencing available to make two identical rectangular calving pens along already-existing L-shaped fencing. Find the dimensions of the calving pens of maximum area.

19. An open rectangular box is to be made by cutting out square corners from a rectangular sheet of metal 16" by 20" and folding up the four edges to form a box. Find the dimensions of the box of maximum volume that can be made in this way.

SEQUENCES

Location in sample syllabus: Year 3, Unit 10

Major standards addressed: Algebra, functions

Objectives: ◆ To write the general term of a sequence
◆ To write the general term for an arithmetic sequence
◆ To write a recursive formula for the general term of a sequence

Sequence A will address level 1, and possibly level 2, of the following lesson activities. Sequence B students will do some exploratory work similar to level 2 activities in addition to activities that introduce recursive definitions of sequences and their application through computers.

Motivating question: A $12 000 loan is to be repaid by making monthly payments of $500 plus 1% of the unpaid balance at the beginning of the month. Make a list of the monthly payments over the twenty-four-month payment period plus the total amount of money repaid.

Level 1 (student-centered)

Determine the pattern and use it to extend the given sequence by writing the next three terms. Then write an explicit formula for the sequence.

1. | n | 1 | 2 | 3 | 4 | 5 |
   |-----|---|---|----|----|----|
   | t_n | 2 | 6 | 12 | 20 | 30 |

2. | n | 1 | 2 | 3 | 4 | 5 |
   |-----|---|---|----|----|-----|
   | t_n | 3 | 9 | 27 | 81 | 243 |

3. $\frac{1}{2}$, $\frac{2}{3}$, $\frac{3}{4}$, $\frac{4}{5}$, $\frac{5}{6}$, $\frac{6}{7}$

4. 11, 17, 23, 29, 35, 41

[1. $t_n = n^2 + n$ 2. $t_n = 3^n$ 3. $t_n = n/(n + 1)$ 4. $t_n = 6n + 5$]

Teaching Matters: Encourage students to verbalize t_n as "the nth term or the term at position n." Suggest that they rewrite each term of the sequence, using the index value as part of the expression before writing the nth term.

Students are given an activity sheet with sequences of figures composed of unit squares. They are asked to use small ceramic tiles or other inexpensive squares to form the figures of each sequence. Hints are given for the first few to furnish ideas of how to manipulate the figures to make the sequence pattern more apparent.

In exercises 5–8, determine a pattern and use it to predict the squares needed to make the eighth figure of the sequence.

5.

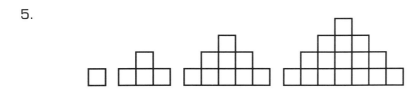

Hint: Move squares in each figure to make a new figure whose shape is simpler.   [64]

*Teaching Matters: **Some of the sequences (such as in exercise 7) will be arithmetic. The formula for an arithmetic sequence can be intuitively developed, leading into other numerical examples of arithmetic sequences and applications of the formula $a_n = a_1 + (n - 1)d$.***

6.

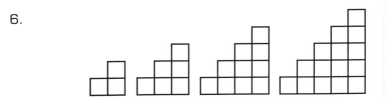

Hint: Try adding an identical shape to each figure to make a simpler shape and then taking half the total squares for the new shape.　[36]

7.

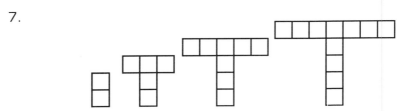

Hint: Make the first figure of the sequence, then transform it into the second, and then the second into the third, and so on. Note the regular change.　[23]

8.

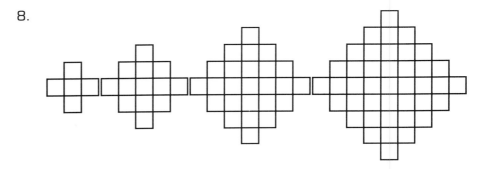

Hint: Make into two simpler shapes by moving squares.　[145]

Level 2

Students are introduced to sequences through numerical examples and geometric examples of the following types.

9. Lines are drawn to intersect all other lines in each diagram, and the points of intersection are counted to form a sequence. Find the next three terms of the sequence and write an expression for the number of intersections for n lines.

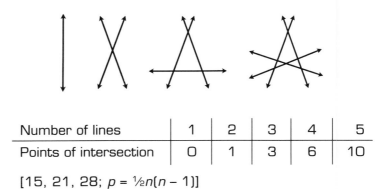

Number of lines	1	2	3	4	5
Points of intersection	0	1	3	6	10

[15, 21, 28; $p = \frac{1}{2}n(n - 1)$]

10. Diagonals are drawn from each vertex of a convex polygon to every other vertex, forming a sequence according to the number of sides of the polygon. Find the next three terms of the sequence and write an expression for the number of diagonals for an n-sided polygon.

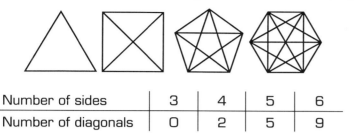

Number of sides	3	4	5	6
Number of diagonals	0	2	5	9

[14, 20, 27; $d = \frac{1}{2}n(n - 3)$]

Level 3

A *recursive definition* of a sequence states the first term of the sequence (and possibly others) and then states how succeeding terms are related to the preceding term(s). Exercise 11 is the recursive definition of an arithmetic sequence. Exercise 12 defines a Fibonacci sequence recursively. Write the first eight terms of each sequence.

11. $t_1 = 10$

$t_{n+1} = t_n + 7$, for $n \geq 1$

[10, 17, 24, 31, 38, 45, 52, 59]

12. $t_1 = 1$

$t_2 = 1$

$t_{n+2} = t_{n+1} + t_n$, for $n \geq 1$

[1, 1, 2, 3, 5, 8, 13, 21]

13. The BASIC program at the right uses the recursive definition to print the first ten terms of the Fibonacci sequence. Run the program on a computer. Modify the program to print 30 terms of a Fibonacci-type sequence whose first two terms are –8 and 5.

```
10 T(1) = 1
20 T(2) = 1
30 PRINT T(1)
40 PRINT T(2)
50 FOR N = 1 TO 8
60 T(N+2) = T(N+1) + T(N)
70 PRINT T(N+2)
80 NEXT N
```

Level 4

Some Sequence B students will be able to do the following activities.

14. In round-robin athletic tournaments, the number of games that must be played according to the number of teams entered is given by the following sequence. Find an explicit formula for this sequence (Hint: Apply the method of finite differences).

Number of teams	1	2	3	4	5	6
Games required	0	1	3	6	10	15

[$G = \frac{1}{2}t(t - 1)$]

15. The number of cannonballs stacked in a three-sided pyramid formation form a sequence according to the number of layers present. Use finite differences to find an explicit formula for the following sequence of cannonball totals according to the layers in the pyramid.

Number of layers	1	2	3	4	5	6
Total cannonballs	1	4	10	20	35	56

[$T = \frac{1}{6}n(n + 1)(n + 2)$]

Teaching Matters: **Following this introductory sequence work, an explicit formula for an arithmetic sequence can be derived and applied to numerical sequences and real-world problems such as the motivating question for this lesson.**

Teaching Matters: **Students should apply both explicit and recursive formulas for sequences to explore applications similar to that of the motivating question of this lesson. Strong use should be made of computers to study real-world applications and to verify answers achieved by formulas. Computer programs illustrate the power of recursive formulas.**

Teaching Matters: **In Unit 1 of Year 3 (see syllabus in Appendix I), students were introduced to the arithmetic application of finite differences to extend patterns of number sequences. In this lesson, students operating at level 4 use the technique of finite differences and systems of equations to determine the explicit formula for sequences where sufficient terms are known to apply the technique. Problems involve sequences whose general term is difficult to find without an analytical technique.**

DEDUCTION WITH VECTORS

Location in sample syllabus: Year 4, Unit 9

Major standard addressed: Geometry from an algebraic perspective

Objective: To write geometric proofs using vectors

Prerequisites: ◆ Equality of vectors: If $\vec{AB} = \vec{CD}$, then $\vec{AB} \parallel \vec{CD}$ and $AB = CD$.

◆ Multiplication by a scalar multiple: If $\vec{QR} = k\vec{ST}$ then $\vec{QR} \parallel \vec{ST}$ and $QR = k(ST)$.

◆ Triangle property of vectors: $\vec{AB} + \vec{BC} = \vec{AC}$

Motivating question: Recall the property of the midline of a triangle: In any triangle, the segment joining the midpoints of two sides is parallel to the third side and half as long. In Year 2, we studied proofs of this property in terms of coordinates, transformations, and Euclidean-type reasoning. Consider how vectors could be used to prove this property by forming a triangle with vectors.

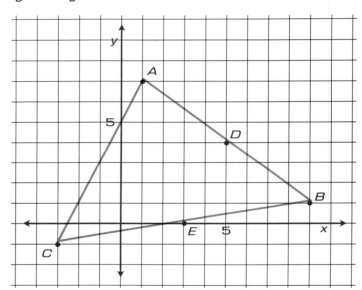

Directions for the initial activity: Consider $\triangle ABC$ with $A(1, 7)$, $B(9, 1)$, and $C(-3, -1)$.

1. Find the coordinates of D, the midpoint of $\overline{AB}$, and E, the midpoint of $\overline{BC}$. [$D(5, 4)$, $E(3, 0)$]

2. Write the vectors for $\vec{DB}$, $\vec{BE}$, and $\vec{DE}$. [$\vec{DB} = (4, -3)$, $\vec{BE} = (-6, -1)$, and $\vec{DE} = (-2, -4)$]

3. Verify that $\vec{DB} + \vec{BE} = \vec{DE}$.

4. Write the vectors for $\vec{AB}$, $\vec{BC}$, and $\vec{AC}$. [$\vec{AB} = (8, -6)$, $\vec{BC} = (-12, -2)$, and $\vec{AC} = (-4, -8)$]

5. Verify that $\vec{BA} + \vec{AC} = \vec{BC}$.

6. Verify that $\vec{DE} = \frac{1}{2}\vec{AC}$.

7. What *two* conclusions can be drawn from the last statement? [$\overline{DE}$ is parallel to $\overline{AC}$ and half as long.]

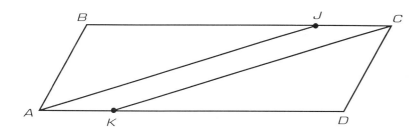

Level 3

[Editorial note: The fourth year is designed to focus on mathematics beyond the core that is needed by college-intending students. Therefore, levels 1 and 2 are omitted in this lesson.]

The example above can be generalized. The reasoning is given below. Explain why each step is true.

8. a. $\overrightarrow{DE} = \overrightarrow{DB} + \overrightarrow{BE}$ [Triangle property of vectors]

   b. $\overrightarrow{DB} = \frac{1}{2}\overrightarrow{AB}$ [M is the midpoint, meaning of scalar multiple]

   c. $\overrightarrow{BE} = \frac{1}{2}\overrightarrow{BC}$ [K is the midpoint, meaning of scalar multiple]

   d. $\overrightarrow{DE} = \frac{1}{2}\overrightarrow{AB} + \frac{1}{2}\overrightarrow{BC}$ [Statements a, b, and c and substitution]

   e. $\overrightarrow{DE} = \frac{1}{2}(\overrightarrow{AB} + \overrightarrow{BC})$ [Distributive property]

   f. $\overrightarrow{AB} + \overrightarrow{BC} = \overrightarrow{AC}$ [Triangle property of vectors]

   g. $\overrightarrow{DE} = \frac{1}{2}\overrightarrow{AC}$ [Statements e and f and substitution]

   h. $\overline{DE} \parallel \overline{AC}$ and $DE = \frac{1}{2}AC$ [Statement g, meaning of scalar multiple]

Teaching Matters: **Have students revisit earlier proofs of statements involving coordinates, transformations, and synthetic methods. Have them construct corresponding vector proofs and then compare the proofs for efficiency, elegance, preference, and convincing argument.**

Follow-up activities:

Use vectors to prove each of the following:

9. Given quadrilateral *ABCD*, form a new quadrilateral by joining the midpoints of the four sides. Prove that the new quadrilateral, *JKLM*, is a parallelogram. [Draw a diagonal, for example, $\overline{AC}$. Use the above results to show that $\overrightarrow{ML} = \frac{1}{2}\overrightarrow{AC}$ and $\overrightarrow{JK} = \frac{1}{2}\overrightarrow{AC}$, so $\overline{ML}$ and $\overline{JK}$ are parallel and congruent, sufficient conditions for *JKLM* to be a parallelogram.]

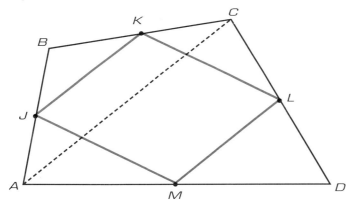

10. If two sides of a quadrilateral are equal and parallel, then the other two sides are also equal and parallel.

11. *ABCD* is a parallelogram. $\overline{JC} = \frac{1}{4}\overline{BC}$ and $\overline{AK} = \frac{1}{4}\overline{AD}$. Prove that *JCKA* is a parallelogram.

Level 4

Vectors and coordinates can be used to make deductions in geometry. Consider a general triangle in the coordinate plane with $A(0, 0)$, $B(2b, 2c)$, $C(2a, 0)$.

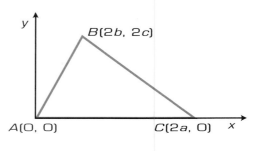

12. a. Find the coordinates of D, the midpoint of $\overline{AB}$, and E, the midpoint of $\overline{BC}$. [$D(b, c)$, $E(a + b, c)$]

   b. Find vectors for $\overrightarrow{AB}$, $\overrightarrow{BC}$, and $\overrightarrow{AC}$. Show $\overrightarrow{AB} + \overrightarrow{BC} = \overrightarrow{AC}$. [$(2b, 2c) + (2a - 2b, -2c) = (2a, 0)$]

   c. Find vectors for $\overrightarrow{DB}$, $\overrightarrow{BE}$, and $\overrightarrow{DE}$. Show that $\overrightarrow{DB} + \overrightarrow{BE} = \overrightarrow{DE}$. [$(b, c) + (-b + a, -c) = (a, 0)$]

   d. Show that $\overrightarrow{DE} = \frac{1}{2}\overrightarrow{AC}$.   [$(a, 0) = \frac{1}{2}(2a, 0)$]

   e. What two conclusions can be drawn from the last statement? [$\overline{DE}$ is parallel to $\overline{AC}$ and half as long.]

RATE OF CHANGE

Location in sample syllabus: Year 4, Unit 10

Major standard addressed: Conceptual underpinnings of calculus

Objectives: ◆ To associate slopes of secant lines with average rates of change

◆ To associate slopes of tangent lines with instantaneous rates of change

Prerequisites: Knowledge of functions, slopes, and quadratic modeling

Materials: Graph paper, graphing utility

Motivating question: A projectile was propelled into the air from a height of 44 feet at a velocity of 80 feet per second. What was its velocity at the instant it hit the ground?   [–96 feet/second]

Directions for the initial activity:

1. Recall that the acceleration due to gravity is –32 feet per second per second; in free fall, it is $-\frac{1}{2}gt^2$. Use this fact to write a formula modeling height as a function of time for the above phenomenon.
   $[h(t) = -16t^2 + 80t + 44]$

2. Graph the function on a graphing utility. Use the graph to answer these questions:

   a. At what time did the projectile reach its maximum height? [2.5 seconds]

   b. What was the maximum height?   [144 feet]

   c. How many seconds was the projectile in the air?   [5.5 seconds]

3. Recall that velocity is change in height divided by change in time.

   a. During what time periods did the projectile have positive velocity? $[0 < t < 2.5$ seconds]

   b. When did it have negative velocity? $[2.5 < t < 5.5$ seconds]

   c. Was the velocity ever zero? Explain. $[t = 2.5$ seconds]

Levels 3 and 4

Consider the average velocity of the projectile in various intervals.

4. Average velocity $= \dfrac{\textit{change in height}}{\textit{change in time}}$.

   For example, consider the interval from the moment of projection $(t = 0, h = 44)$ to the moment the projectile reaches its maximum height $(t = 2.5, h = 144)$. Its average velocity during this period was

   $$\frac{\text{average}}{\text{velocity}} = \frac{\textit{change in height}}{\textit{change in time}} = \frac{144 - 44}{2.5 - 0} = 40 \text{ feet per second.}$$

   a. Draw a line on the graph joining these two points. Such a line intersecting a curve in two points is called a *secant line.*

   b. Find the slope of the line and compare this value to the average velocity. Explain your observation.   [Slope of line = average velocity]

c. Draw the secant line determined by the points corresponding to $t = 2$ and $t = 4$ seconds. Estimate the average velocity for this interval. [Answers will vary but should be close to -16 ft./s.]

d. Determine the heights at $t = 2$ and $t = 4$; then find the average velocity for that interval. [At $t = 2$, $h = 140$ ft.; at $t = 4$, $h = 108$ ft.; average velocity $= -16$ ft./s]

e. Write a general formula for the average velocity between $t = a$ and $t = b$, assuming a < b. Use function notation for the height.

$$\left[\text{Average velocity} = \frac{h(b) - h(a)}{b - a} \right]$$

5. We can estimate the instantaneous velocity of the projectile as it hits the ground by looking at smaller and smaller average velocities in times close to $t = 5.5$ seconds. Find the following average velocities:

a. between $t = 3$ and $t = 5.5$ seconds
[Answers will vary; about -56 ft./s]

b. between $t = 4$ and $t = 5.5$ seconds
[Answers will vary; about -72 ft./s]

c. between $t = 5$ and $t = 5.5$ seconds
[Answers will vary; about -88 ft./s]

d. between $t = 5.25$ and $t = 5.5$ seconds
[Answers will vary about -92 ft./s]

Describe and estimate an *average* velocity at $t = 5.5$ seconds.

6. Just as the slope of a secant line gives the average velocity for an interval, the slope of the tangent line gives the instantaneous velocity at a point.

a. Sketch tangent lines to the graph of the projectile at the following times. Estimate the slopes of the tangent lines.

$$t = 1 \qquad t = 2.5 \qquad t = 4 \qquad t = 5$$

[48 ft./s, 0 ft./s, -48 ft./s, -80 ft./s]

b. Another way to estimate instantaneous velocities is to zoom in on the image of the graph until the curve is virtually a line. The slope of this "line" gives a good estimate of the slope of the tangent line to the curve at this point. Use the zooming feature of the graphing utility to estimate the slope of the curve for the above times. Compare your results with those of other students.

c. Create a graph of the function near the point (5.5, 0). Include values below the x-axis. Use the zoom procedure to estimate the slope of the curve near this point. Compare your results to those in exercise 5d.

7. Sketch tangent lines along the function at every 0.5-second interval. Write a sentence or two to explain how the slopes of these tangent lines change as time changes from 0 to 5.5 seconds.

[Slopes of tangent lines are positive and steep, growing less steep until the slope becomes 0 at $t = 2.5$; then slopes are shallow and negative, growing more steep until $t = 5.5$.]

8. We have seen one form for the average velocity (see exercise 4e). Another form is shown and explained below. The two forms are equivalent.

$$\frac{f(b) - f(a)}{b - a} = \frac{f(a + \Delta x) - f(a)}{\Delta x}$$

When considering changes in the time variable, we sometimes let $\Delta x = b - a$. This is read "delta x," or the difference in the time equals the difference of the x-values at b and a. Let a be the fixed point. As b changes, Δx changes. Compare these two diagrams and explain why they are equivalent. [The value b has been re-expressed as $a + \Delta x.$]

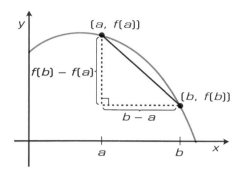
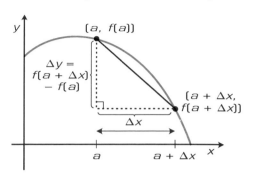

Follow-up activities:

Most functions have a special relationship to their instantaneous rate-of-change function. Some of these relationships are easier to recognize than others. Have students work in pairs to investigate the following functions. In each case, they should find the slope of the tangent (instantaneous rate of change) at several points along the curve. They may use either the estimated tangent line or the zoom procedure. Have each pair of students make a table of their findings and describe whatever pattern they observe.

9. $f(x) = 0.5x^2$ 10. $g(x) = e^x$ 11. $h(x) = \frac{1}{3}x^3$ 12. $k(x) = -x$

[9. Slope equals x; 10. slope equals the value of the function; 11. slope equals x squared; 12. slope is -1 everywhere]

*Teaching Matters: **Students should do comparable activities with other functions. Continue to ask questions like that of exercise 3 of the initial activity and exercise 4c in which students are directed to look at graphs, judge whether changes are positive or negative, and estimate slopes. With appropriate software, students can do the entire exploration on a computer screen. When using the zooming procedure of exercise 6b, have students zoom in several times to increase their accuracy. In exercise 8, the notation uses a and b to avoid subscripts. After students are comfortable with the second expression for average change, you may want to replace a with x.***

*Assessment Matters: **The activities in these prototype lessons suggest possible test items, open-ended questions, long-term projects, and group investigations that can be drawn on for assessing students.***

CHAPTER 4
ENRICHMENT AND DIFFERENTIATED CURRICULUM MODELS

The core curriculum seeks to provide a fresh approach to mathematics for all students—one that builds on what students can do rather than on what they cannot do.
—Curriculum and Evaluation Standards

This chapter presents two more models for a core curriculum—the Enrichment Model and the Differentiated Model. Both are single-sequence course models that use different mechanisms for adjusting instruction to student differences. Both are unified approaches to a 9–12 program, fully integrating the treatment of algebra, geometry, statistics, and discrete mathematics. They represent robust versions of a core curriculum, making extensive use of technology, blending a rich mixture of topics from many areas of mathematics, and illustrating the most effective development of connections among mathematical ideas.

THE ENRICHMENT MODEL

The Enrichment Model consists of the same core material to be completed by all students, with additional, carefully selected activities for student exploration. The model takes its name from the fact that some students will enrich their experience with the core topics by further study. Each unit of instruction consists of a sequence of activities ranging from concrete to abstract. Students progress in small groups through a variety of activities focused on a particular content area. Some groups (or individuals) may move more quickly through a unit than others and therefore complete additional work at higher levels of abstraction. After the allotted time period (e.g., two to three weeks) for the unit, the entire class progresses to the next unit.

The first three courses constitute the core, and the fourth course provides the additional mathematics needed by college-intending students. The Enrichment Model is an option for accommodating students who move faster through the core material without denying access to the core to those students who move at a slower pace. Since it could be implemented within a single class of students, it is a viable alternative for schools of any size. (See fig. 4.1.)

Features of the Model

A key element of the Enrichment Model is its flexibility. The version outlined below furnishes enrichment options for each of the ten or more units covered each year in the core curriculum. When the model is implemented in a single classroom, students have the flexibility to enrich their experiences, or not, every few weeks according to their progress through the core material. The choice they make for one topic can be quite independent of the choice made for another. Since each enrichment lesson stands alone, there is usually little difficulty posed by lack of exposure to prerequisite material.

Another advantage of the model is student comfort level. Students who do not participate in an enrichment activity should feel less pressure in studying the core ideas to move beyond their pace for understanding because of peer pressure or a fear of holding others back. While some classmates are studying the enrichment units, they can feel more comfortable asking questions and delving into ideas with

Students are regrouped at the beginning of each unit of core curriculum topics (rectangles). Students will need varying amounts of time for investigating and solidifying their understanding of these topics. Students who complete the unit in less time will move laterally with regrouping to the study of enrichment material (E). Differentiation is accomplished by the effective use of technology and classroom management strategies.

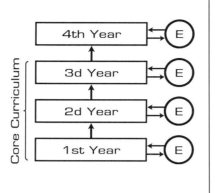

Fig. 4.1. Enrichment Curriculum Model. The Enrichment Model is a single sequence of courses with the core curriculum comprising the first three years. Enrichment topics provide for horizontal enrichment on a unit-by-unit basis.

classmates who are also taking more time with the core material. This might be especially beneficial for students who are confident of learning the material but who are not as quick to express themselves in the whole class. The enrichment feature can also provide for students whose personal circumstances are changing. For example, a student who usually participates in the enrichment units and who is ill for a week would not need to make up the enrichment material to keep up with the new material in the core.

Students who do participate in enrichment activities will gain opportunities beyond those typically possible in conventional classes. They will be able to look more deeply into topics, to make more challenging explorations, and to make broader connections within and outside of mathematics. They will also have more freedom to pursue individual interests, such as doing research. These additional opportunities for studying interesting topics furnish an incentive for students to complete the core material in a timely manner.

Instructional Considerations

The use of enrichment within a single class is facilitated by enrichment materials that can stand alone. Students doing enrichment units will at times work independently, but it is important that their work in the enrichment modules be carefully monitored. Connections among the enrichment and the core topics can be better made by dialogs between teacher and students and by presentations of topics to the whole class by those engaged in enrichment.

Enrichment topics can fall into one or more of the following categories:

Historical and cultural topics—taking more time to delve into the history and cultural contributions of topics such as right triangle trigonometry or ornamental patterns

Abstraction and generalization—addressing topics at a more abstract level, such as calculating the lengths of the diagonals of a hypercube with O-1 coordinates, considering functions of several variables when studying functions of one variable, or probing the study of isomorphisms or affine transformations

New settings—looking at concepts in fresh settings, such as investigating geometric concepts in finite geometries, solving Diophantine equations, or looking at the applications of mathematics in careers

TI-81 program to plot a fractal using random numbers:

```
PrgmF:FRACTAL
2 → X
1.5 → Y
Lbl 4
Int(Rand*3)+1 → A
If A=1
Goto 1
If A=2
Goto 2
If A=3
Goto 3
Lbl 1
(X+1)/2 → X
(Y+1)/2 → Y
Pt-On(X,Y)
Goto 4
Lbl 2
(X+3)/2 → X
(Y+1)/2 → Y
Pt-On(X,Y)
Goto 4
Lbl 3
(X+2)/2 → X
(Y+(1+√3))/2 → Y
Pt-On(X,Y)
Goto 4
```

Note: Range of graphics screen should be set at [1, 3] × [1, 3].

New perspectives—gaining new perspectives and insights, such as investigating fixed-point algorithms for solving equations or curve fitting reciprocals after doing a transformation

Deeper explorations—exploring relationships more deeply or exploring new relationships beyond those in the core, such as studying paradoxes in a topic area, exploring unsolved problems, exploring number theory, or studying specific topics such as the constructibility of squares on a geoboard, the generation of random numbers, or the properties of combined functions such as $y = x \sin x$ and $y = e^{\sin x}$

Further examples of connections—making connections such as relating set properties and logic by introducing DeMorgan's laws or relating geometry and computation by introducing geometric constructions for addition, subtraction, multiplication, and division of real numbers

Enrichment Possibilities

An enrichment approach could be employed with the Crossover and other models for a core curriculum. But for purposes of presentation, it is aligned with the sample syllabus for the Differentiated Model found in Appendix II (a four-year unified mathematics curriculum).

SAMPLE ENRICHMENT TOPICS

Year 1

Unit	Possible enrichment topic	Relating to... in unit	Comments
Exploring (plane) geometric figures	Finite geometries	Basic vocabulary and concepts	Investigate concepts such as parallel lines in finite geometries, where the concepts will be analogous but will look different.
Exploring data	Representing geographic data by scale drawings; creating databases for future analysis	Representing data	A map of Africa can be redrawn to illustrate each country as a function of its population. If 1 cm² represents 1 million people, Nigeria would have an area of 115 cm² and Malawi an area of 8 cm².
Graphs	Finite graphs using modular numbers	Representations of solutions of open sentences	Relate the graphs of equations over the integers modulo 3. For example, the graph of $x + 2y = 1$ (mod 3) will consist of three points with coordinates (0, 2), (1, 0), and (2, 1), and the graphs of $x + 2y = 0$ and $x + 2y = 2$ will be lines that are "parallel."
Expressions, sentences	Tautologies and inconsistencies	Equivalent and nonequivalent expressions	Identify sentences that are always true and their proofs.
Models for operations	Geometric (length) models for sums, differences, products, and quotients	Models for operations	This could be done using geometric software (such as the Geometer's Sketchpad).
Linear situations, sentences, graphs	Diophantine equations	Solving linear equations	This could include problems like "the chickens and cows have 50 legs and 16 heads" problem. (How many of each are there?)

Unit	Possible enrichment topic	Relating to... in unit	Comments
Special powers	Generating Pythagorean triples and Fermat's last theorem	Pythagorean theorem	
Properties of geometric figures	Fractals	Perimeter and area	Investigate fractal dimension using guess-and-check (e.g., to solve $3^D = 4$, use the calculator exponent key to estimate that D is about 1.26).
Measures in geometry	Archimedes' approach to area and volume, such as the area under a parabola	Areas of irregular figures	
Probability and simulation	Fractals	Simulation	See the sidebar on p. 68 for a program that generates random ordered pairs of numbers, which when plotted result in a fractal.
Functions	Graphing relations as unions of functions and solving problems about relations	Solving equations using a function grapher	For example, solve the system $x^2 + y^2 = 4$ and $y = x^3$ by graphing the circle as $y = \pm\sqrt{4 - x^2}$.

Year 2

Unit	Possible enrichment topic	Relating to... in unit	Comments
Variation and modeling	Chaos theory	Introduction to modeling	See the use of quadratic functions to model animal populations in *Chaos* by James Gleick.
Coordinate geometry	Constructibility of figures on a lattice (geoboard)	Proofs in coordinate geometry	For example, what size squares can be constructed on a geoboard?
Transformations and geometric figures	Symmetry groups, such as that of an equilateral triangle	Composition of trans-formations	Possibly include expressing translations and rotations as compositions of reflections.
Introduction to trigonometry	The historical perspective on the geometric interpre-tation of the names of the trigonometric functions	Definitions of trigo-nometric functions	For example, define the "tangent" as the length of a tangent segment to a unit circle, and the "cotangent" as the length of a tangent generated by the complementary angle.
Functions	Functions of several variables	Functions of one variable	For example, find and optimize the volume of a box with dimensions l, w, and h under given constraints.
Lines, parabolas, and exponential curves	The role of e in mathematics, natural sciences, and business	Exponential functions	
Transformations of functions and data	Generalizing the effect of transformations on data, including proofs	Transformations of data	For example, replace x-values by $x + b$, ax, or $ax + b$ and investigate the effect on measures of central tendency and dispersion. This can be done more generally than in the regular material.

Unit	Possible enrichment topic	Relating to... in unit	Comments
Systems	Graphing in three dimensions	Graphing systems	For example, what is the graph of $2x + 3y - z = 12$?
Matrices	Matrix representations for transformations and their products	Interpreting products of matrices	As the core treats matrices more informally, this "traditional" approach becomes enrichment.
Combinatorics and binomial distributions	Multinomial coefficients	Binomial coefficients and Pascal's triangle	For example, Pascal's triangle can be used to solve this problem: How many ways are there to pick a committee of 6, with at least one member from each high school grade?

Year 3

Unit	Possible enrichment topic	Relating to... in unit	Comments
Fitting curves to data	Fitting reciprocals (and other functions)	Curve fitting	If the data seem to fit a curve of the form $y = k/x$, replace x by $X = 1/x$ in the data points and find a linear fit of these points, which can be used to find k in $y = k/x$. For example, this method can be used to find $k = 72$ in the rule of 72.
Circular functions and models	Graphs of sums and products of linear and trigonometric functions	Graphs of trigonometric functions	For example, graph $y = x + \sin x$ and $y = x \sin x$.
Exponential and log functions	Curve fitting exponentials and log functions using log transformations	Log transformations (and chapter 1 curve fitting)	For example, fit $y = ab^x$, as $\ln y = \ln a + (\ln b) x$.
Logic	DeMorgan's laws (relating to sets)	Quantifiers and negations	
Reasoning in geometry	Fermat's optimization problem to find a point that minimizes the sum of the distances from three given vertices of an acute triangle.	Geometric models and transformations	This can be presented as "the cable TV problem," finding where to locate a cable TV station so that the sum of the distances from three cities is minimal. Have students find solutions by guess-and-check, using software such as the Geometer's Sketchpad. Then show a solution that has an elegant, yet accessible proof (which can be found in Coxeter's *Introduction to Geometry* [1969]) to illustrate the power of proof.
Reasoning in algebra	Further examples and generalizations of isomorphisms	Finite systems and isomorphisms	
Reasoning in intuitive calculus	Contrasting methods to find areas under curves (rectangles, trapezoids, Simpson's rule)	Area under curves	This unit makes ample use of technology.

Unit	Possible enrichment topic	Relating to... in unit	Comments
Reasoning in discrete mathematics	Fibonacci searches	Systematic searches	
Reasoning in probability	Bayes' theorem	Conditional probability	
Reasoning in statistics	Paradoxes in statistics	Statistical reasoning	Investigate Simpson's paradox (e.g., a baseball player's average against right-handed pitchers and his average against left-handed pitchers are both higher than another player's, yet his overall average is lower).

THE DIFFERENTIATED MODEL

The Differentiated Model consists of a single-course sequence in which topics are addressed at various levels of concreteness and abstraction. The first three courses constitute the core and the fourth course provides additional content for the college-intending student. (Fourth-year options could be designed that also address specialized applications that support the technological workplace.) The core curriculum is embedded within each of the first three years, with all students studying all topics. Differentiation of learning outcomes occurs according to the level(s) of activities students complete for each topic. All four years consist of a fully content-integrated program that assumes a K–8 preparation consistent with the K–8 *Curriculum and Evaluation Standards* (NCTM 1989). (See fig. 4.2.)

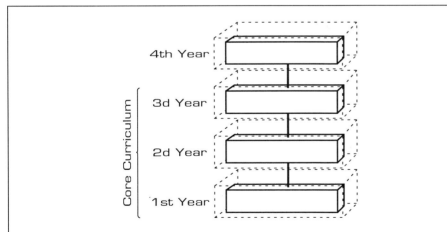

Classes are heterogeneous, but students are frequently regrouped to put together students who are working at similar levels with respect to a particular topic. Some students will complete levels 1–3 of a given topic, others will complete levels 1–4, and some will complete levels 3–5 or 4–5. Design of materials, extensive use of technology, and innovative classroom management and methodology allow differentiation of learning to occur.

Fig. 4.2. Differentiated Curriculum Model. The solid parallelepipeds represent the core level of outcomes for all courses. The dashed parallelepipeds represent levels of outcomes that exceed the core. All topics are designed to incorporate several differentiated levels of the intended learning outcomes.

Features of the Model

A possible syllabus for the Differentiated Model is provided in Appendix II. It attempts to weave the goals of standards 1–4 of the NCTM grades 9–12 standards into every course and the mathematics of the other standards into most courses. All courses feature student investigations of topics centered in real-life situations when possible. Students are expected to be *active* learners, using technology as appropriate and writing and talking about (as well as listening and reading about) ideas they are discovering and applying. Each year the syllabus covers content from four strands: algebra/functions, geometry/trigonometry, statistics/probability, and discrete mathematics/calculus underpinnings. We have elected to use a maximum of twelve units in a year to provide a program that is, in fact, doable in the reality of most school settings.

Some features of the model are not obvious from the notes in Appendix II. For one, we assume that review is spiraled from lesson to lesson, unit to unit, and course to course. Student assignments will regularly include items from earlier studies. Also, the curriculum takes advantage of available and emerging technology. We assume the availability of graphing utilities, geometry exploration software, software for creating data displays and calculating statistics, and symbol manipulation software.

We consider the syllabus for this model to be transitional—that is, not fully realizing the objectives of the *Curriculum and Evaluation Standards* but moving very much toward them. In a more "robust" syllabus, we would expect that unit titles and topics would look even more unified. Lessons might stem from rich real-world situations that embed a variety of mathematical learnings in a mix not so segmented into traditional titles. And a good portion of the first year, in a future revision, might be expected to be studied before grade 9.

Instructional Considerations

Where does the space or the time come from to put so much new material in the sequence? One big reduction is in the elimination of a separate year of geometry. Although geometric (or trigonometric) ideas occur in each course, the heavy emphasis on two-column synthetic proof is gone. Instead, students explore geometry in a more eclectic fashion, taking advantage of more connections in the curriculum. In the first year, geometry ideas are approached inductively through paper folding, constructions, measurement, and deductions stemming from unquestioned assumptions.

In the second year, students discover and prove further properties by using coordinate methods. Through the study of transformations, students develop a general understanding of congruence and similarity. This approach serves many purposes. It connects geometry with algebra, and later with functions, so that translations and scale changes of functions can be easily understood. It provides a general definition of congruence and similarity that can be generalized to all figures in the plane and in space. Using variable coordinates in analytic proofs develops from a base in which students use numeric coordinates, helping to build the idea of arguing from a general case and distinguishing deductive from inductive reasoning. In this second year, students learn the trigonometric ratios for right and oblique triangles. This choice allows them to explore the uniqueness of triangles in terms of triples of information (e.g., *SAS*). Consequently, sufficient conditions for congruent triangles are addressed through the concrete avenues of construction and verification with algebra. Therefore, these topics do not need to be addressed independently of trigonometry.

Another strategy for "making space" in the curriculum is reducing emphasis on algebraic manipulations, particularly factoring and working with rational expressions. Our axiom is that concepts are more powerful than procedures and more accessible to more students. Students can, if necessary, use symbol manipulators for more complex algebraic work, just as elementary students use calculators.

A third major strategy has been to dovetail topics. Triangle congruence is taught from a trigonometric, rather than a synthetic geometry, perspective. Modeling, particularly finding an equation to approximate a set of data, is coupled with data analysis and functions. Spreadsheets produce output recursively, blending discrete processes with explorations in statistics, functions, and algebra.

Differentiation in learning outcomes occurs by blending core lessons for all students with extended activities that students can complete to different levels of abstraction. Each lesson begins with a core activity ideally launched from some significant question that the new concepts address. The initial activity is generally at a concrete or semiconcrete level. This activity is followed by questions and discussion leading to key ideas and applications. Following the discussion and activities, students are offered a sequence of follow-up activities that build from one level to the next. These are at levels 2 and 3, and possibly at levels 4 and 5, using the same "level" ideas as illustrated in the *Curriculum and Evaluation Standards* and further elaborated in Hirsch and Schoen (1989). All activities can be done by individuals, student pairs, or small groups. We picture one to two days spent on an initial class activity and two to four days on the follow-up activities. Follow-up activities culminate by having students discuss what they have learned. All activities and discussions include teacher observation, inquiry into student findings and reasoning, intervention, and clarification.

Benefits

We see many benefits of this model.

♦ By starting all students together on a basic core activity and discussion, teachers can be sure that the foundation ideas of a lesson are addressed for everyone. If necessary, individuals or groups who need further exploration at this level can do so while other groups do follow-up activities.

♦ Nesting follow-up activities (e.g., level 3 subsumes level 2) allows individuals or groups to move ahead to their own highest level of abstraction for a given lesson. We recognize that students may differ in the level of abstraction they can achieve in each lesson. Thus we avoid the traditional tracking problem that locks students into a course at one level of abstraction.

♦ By delaying until Year 4 the objectives for the college-intending student, *late bloomers* have the maximum opportunity to choose this path. This curriculum feature promotes full realization of the intent of the core curriculum: maximizing learning opportunities for all students.

♦ The concrete, reality-based approach to learning activities, capitalizing on technology, gives this curriculum the potential to be truly accessible to a wide spectrum of secondary school students.

PROPERTIES OF BISECTORS

> *Location in sample syllabus:* Year 1, Unit 1
>
> *Major standard addressed:* Geometry from a synthetic perspective
>
> *Objectives:* ◆ To recognize, justify, and apply—
> - properties of angle bisector
> - properties of perpendicular bisector
>
> *Prerequisites:* Vocabulary: *bisect, perpendicular, angle measure, circle, radius, ray, rhombus*
>
> *Concept:* Diagonals of a rhombus are perpendicular bisectors of each other
>
> *Materials:* Compasses, rulers, protractors

Teaching Matters: **This situation should be discussed with students so that they appreciate the concern for finding a location that meets a number of conditions (locus), understand the conditions given, and develop some sense of what might constitute a solution. Point out that the following initial activity addresses some of the related, simpler questions that lead to a full solution of the motivating question.**

Motivating question:

An emergency heliport is to be built so that it will be equidistant from the centers of three towns. How should its location be determined?

a. What are the conditions?

b. Draw a diagram.

c. Could you change one or more conditions to create a simpler problem?

Directions for the initial activity:

Level 1

1. a. Draw a segment, $\overline{XY}$.

   b. Roll your paper so that X coincides with Y. Crease.

   c. Unfold the paper and label the point O where the fold intersects $\overline{XY}$. Label another point on the fold W.

   d. Measure $\angle WOX$ and $\angle WOY$.   [Each measures 90°.]

   e. Measure $\overline{OX}$ and $\overline{OY}$. Compare.   [They are equal.]

   f. Measure $\overline{WX}$ and $\overline{WY}$. Compare.   [They are equal.]

   g. Mark three other points, Q, R, and S, on line WO.

   h. For each point Q, R, and S, find its distance to X and to Y. Compare these two distances for each point.

   i. State your observations.   [All points on OW are equidistant from X and Y.]

Teaching Matters: **Have students share results and state observations from exercise 1. They should recognize that the fold line—**

- **is perpendicular to the original segment;**
- **bisects the original segment;**
- **is a set of points equidistant from the endpoints of the segment.**

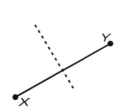
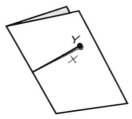
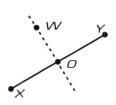

2. Mai walks from the school building to the road. She wants to find the shortest path. How should she walk? [Along the perpendicular from the school doorway to the road]

3. a. Draw an obtuse angle, $\angle AOB$, and find its measure.

   b. Roll your paper so that rays OA and OB coincide. Crease.

   c. Unfold and label a point C on the fold in the interior of $\angle AOB$.

   d. Measure $\angle AOC$ and $\angle COB$. Compare. [The angles are equal.]

   e. Repeat steps a–d with an acute angle.

   f. State a conjecture. [The fold produced by aligning the rays of an angle bisects the angle.]

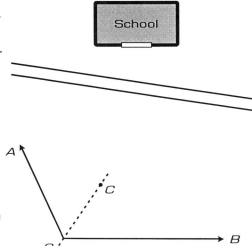

Teaching Matters: **Discuss exercise 2. Draw several possible segments and note that only one segment is perpendicular; it is also the shortest segment. Have students justify this conclusion. [The road, any other path, and the perpendicular path form a right triangle. Any other path will always be the hypotenuse and therefore longer.] Point out that we define the distance from a point to a line as the length of the perpendicular to that line.**

Discuss the results of exercise 3. Students should recognize that the fold bisects the original angle.

Class discussion:

4. Have students justify their conclusions. Have them refold $\overline{XY}$ on the paper used in exercise 1.

   a. Why must the fold be perpendicular to the segment? [The angles are equal and supplementary.]

   b. Why must the fold bisect the segment? [X and Y must coincide.]

   c. With the paper folded, hold it up to the light and view $\overline{WX}$ and $\overline{WY}$. Why must $WX = WY$? [They coincide.]

5. Point out to students they have informally demonstrated that the perpendicular bisector of a segment consists of points equidistant from the endpoints of the segment. Ask if it is also true that points equidistant from the endpoints must create a perpendicular bisector. Have students perform the following experiment to investigate:

   a. Draw a segment AB. Open the compass to a distance more than half of $\overline{AB}$.

   b. Draw two circles with the same radius, one with a center at A and one with a center at B.

   c. True or false: The circles intersect at points equidistant from the endpoints of the segment. [True]

   d. Repeat steps a–c for at least three different radii.

   e. What appears to be true of all the intersection points? [They lie on the perpendicular bisector of the segment.]

6. Discuss the construction activity and have students give justifications.

   a. Call the points of intersection of a pair of circles C and D. Why must $AC = AD = BC = BD$? [All are radii of the same circle.]

   b. What sort of quadrilateral is $ACBD$? [A rhombus, because all four sides are congruent]

   c. What relationships do $\overline{AB}$ and $\overline{CD}$ have in quadrilateral $ACBD$? [They are diagonals. Diagonals of a rhombus are perpendicular bisectors of one another.]

Teaching Matters: **Discuss the results to exercise 7. Students should observe that the angle bisector consists of points equidistant from the sides of an angle.**

7. Guide students through the following experiment concerning properties of the bisector of an angle:

   a. Draw an angle on a piece of paper and fold so the rays coincide; locate any point P on the fold. Keep the angle folded.

   b. Make a fold through P so the rays of the angle are aligned. The new fold should be through P and perpendicular to each of the two sides of the angle. Why? [Folding a ray or line onto itself forms two congruent, supplementary angles, so each must be a right angle.] This is the distance from P to the rays of the angle. Unfold all folds.

   c. What must be true about the two lengths from P to the sides of the angle? [The distances from P to either side are equal.]

   d. Repeat a–c for other points Q, R, and S.

   e. State a conjecture. [Any point on the angle bisector is equidistant from the sides of the angle.]

8. Have students answer the motivating question now or after the follow-up activities. [Students should recognize that the required location is the intersection of the perpendicular bisectors of two (or three) of the sides of the triangle formed by the centers of the three towns.]

Follow-up activities:

Level 2

9. a. Draw a triangle, $\triangle ABC$. Construct the perpendicular bisectors of $\overline{AB}$ and $\overline{BC}$. Call the intersection point D.

   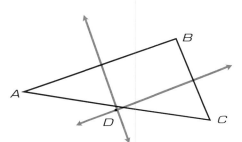

   b. Explain why $AD = BD$. [D is on the perpendicular bisector.]

   c. Explain why $BD = CD$. [D is on the perpendicular bisector.]

   d. Why must $AD = CD$? [Both equal BD.]

   e. True or false: D is on the perpendicular bisector of $\overline{AC}$. Explain. [True; a point equidistant from the endpoints of a segment is on the perpendicular bisector of the segment.]

   f. Draw a circle with center D and radius DA.

   g. What is the relationship of $\triangle ABC$ and circle D? [$\triangle ABC$ is inscribed in D.]

   h. Repeat steps a–g with another triangle.

   i. State a conclusion. [The perpendicular bisectors of the three sides of a triangle intersect in one point. That point is the center of a circle that contains the three vertices of the triangle. The center is equidistant from the vertices of the triangle.]

   j. Look up "circumscribe" and "circumcenter" and relate them to this activity.

10. a. Draw a triangle, $\triangle ABC$. Construct the angle bisectors of $\angle A$ and $\angle B$. Call the intersection point D.

    b. Find the distance from D to $\overline{CA}$, D to $\overline{AB}$, and D to $\overline{BC}$.

    c. Why must the first two distances be equal? [D is on the angle bisector of $\angle A$.]

    d. Why must the last two distances be equal? [D is on the angle bisector of $\angle B$.]

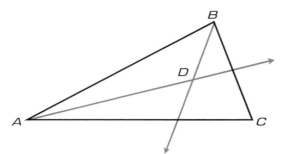

e. Why must the first and third distances be equal? [They each equal the distance from D to AB.]

f. True or false: D is on the angle bisector of $\angle C$. Explain. [True; a point equidistant from two intersecting rays is on the bisector of the angle determined by the rays.]

g. Draw a circle with center D and radius equal to the distance from D to any of the other three sides.

h. What is the relationship of $\triangle ABC$ and circle D? [Circle D is tangent to the three sides of $\triangle ABC$.]

i. Repeat steps a–h with another triangle.

j. State a conclusion. [The angle bisectors of the three angles of a triangle intersect in one point. That point is the center of a circle that contains exactly one point from each side of the triangle. The center is the shortest distance to all three sides.]

k. Look up "inscribe" and "incenter" and relate them to this activity.

11. A camp clinic is to be built so that it will be equidistant from the three roads running through the camp. None of the roads are parallel. Describe how to find this location. [Call the triangle formed ABC. Bisect any two of its angles. The point of intersection of the bisectors is the needed location.]

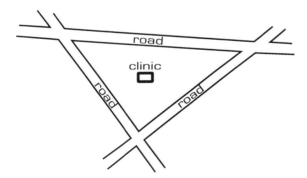

12. Two runners, one at point A and one at point B, are to race toward a road. Where should the finish of the race be located so that the two runners race the same distance? [Where the perpendicular bisector of $\overline{AB}$ intersects the road]

$A \bullet$

$B \bullet$

Teaching Matters: **Perpendicular bisectors coincide outside the triangle if the triangle is obtuse. They coincide at the midpoint of the hypotenuse if the triangle is a right triangle. They coincide inside the triangle if the triangle is acute. Angle bisectors always coincide at a point in the interior of a triangle.**

Teaching Matters: **If two roads are parallel, there are two possible points equidistant from the three roads. At each point, bisect the two interior angles formed by the parallels and the nonparallel road (transversal). If three roads are parallel, there is no point equidistant from them all.**

Teaching Matters: **Locus is concrete, verifiable by measurement, of historic interest, and useful. It provides many opportunities for geometric modeling of real-world phenomena. The suggested activities ask students to explain and justify frequently, promoting the communication and reasoning goals of the Curriculum and Evaluation Standards. Activities use minimal symbolism and delay introducing vocabulary until the need arises. With some background in reflections and symmetry, the examples above and other paper-folding demonstrations can be given a more formal development.**

Level 3

13. Repeat exercises 9 and 10, using geometric construction software. Repeat the directions with triangles that are scalene, isosceles, and equilateral. Then repeat with triangles that are obtuse, acute, and right. Are the intersection points ever on the triangle? Outside the triangle? Under what conditions? What other patterns do you find in the results?

Level 4

14. Reconsider exercise 11. What if two roads are parallel? What if all three roads are parallel?

15. Reconsider exercise 12. Alter the conditions to make the question more challenging. Try to solve your new question.

16. Write and solve other questions that use loci.

PATTERNS TO VARIABLES

Location in sample syllabus: Year 1, Unit 1

Major standard addressed: Algebra

Objective: To recognize a pattern, giving specific instances of the generalization both verbally and symbolically

Prerequisites: Concepts of polygon and perimeter

Materials: Pattern blocks for equilateral triangles, regular hexagons, and isosceles trapezoids (one side length double the other three)

Motivating question: A car rental costs $28.00 a day and $0.45 a mile. What are some possible questions we can answer for this situation?

Directions for the initial activity:

Level 1

1. Choose a polygon from those available from your teacher and find its perimeter.

2. Build chains with your shape as shown below. Each time you add another polygon, find the perimeter of the new figure.

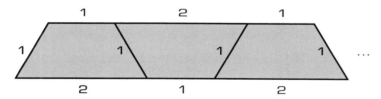

3. Complete this table:

Polygon type: _____							
Number of polygons	1	2	3	4	5	6	7
Perimeter of chain figure							

4. Determine the perimeter of a 10-polygon figure. Check by building it.

5. How would you find the perimeter of a 100-polygon figure?

6. How would you find the perimeter for a figure made from some other number of polygons?

Class discussion:

♦ Have students compare descriptions of their chain figures for one shape. Identify many possible different verbalizations of the patterns. For example, typical student responses for the trapezoid include the following:

"Three times the number of trapezoids plus 2."
"The horizontal sides give 2 plus 1 for each trapezoid, then you add 2 for the slanting edges."
"Triple the number of trapezoids, then add 2."
"Take 2, then add 3 for each trapezoid."

Teaching Matters: Possible responses: What does it cost if you drive 256 miles? What does it cost for three days with mileages of 85, 218, and 73? Is it possible for a bill (without tax) to be $114.40 for one day? How many miles can you go if you can afford $200.00? What number of miles takes your cost for one day over $100.00? What is the procedure for finding the cost for one day's rental?

Tell students that they will be able to answer questions like these after this and other related lessons. In this lesson, the focus will be on finding a general relationship between two values such as mileage and the cost for renting a car for one day.

"2 for the lengths of the right and left edges, plus 3 for the parallel sides of each trapezoid."

"Multiply by 4, then subtract 2 fewer than the number of trapezoids."

- Have students repeat exercises 1–6 for other shapes.
- Ask students to suggest symbols to abbreviate their patterns. Discuss the meaning of the word *variable.*
- Have students graph their information as ordered pairs—the number of polygons on the horizontal axis and the perimeter on the vertical axis. Ask, "By how many units does the second value (perimeter) increase when the first value (number of polygons) increases by 1?" [3 in the case of trapezoids. Relate this to the geometric situation: there are three units on the horizontal edges that get added each time a new trapezoid is included.]

Follow-up activities:

Level 2

For each situation in exercises 7–12, do the following:

a. Identify two variables that are changing.

b. Make a table of values for at least six pairs of these related values.

c. State verbally a general rule that describes any pattern.

d. Write a formula using variables for the pattern, stating what your variables represent.

e. Graph the values from your table and explain (1) how the second value changes as the first increases and (2) this relationship in terms of the situation.

7. Book fines are 10 cents each day the book is late.

8. A rental car for one day costs $28 plus 45 cents a mile.

9. A person weighed 204 pounds and lost 3 pounds a month.

10. The number of sugar cubes needed to build a cube with sides of length 1, 2, 3,...

11. The surface area of a sugar cube with sides of length 1, 2, 3,...

12. Each tape costs $5.98 plus 4 percent sales tax.

13–18. For each situation in exercises 7–12, write two problems that fit the pattern. The two problems should require the reader to find the values for the two variables involved. For example, in the initial activity with trapezoids, you could ask,

a. What is the perimeter for a chain of 234 trapezoids?

b. What number of trapezoids will produce a chain with a perimeter of 305?

19. Find the solution to questions a and b above.   [704, 101]

Level 3

20. Use a spreadsheet to generate a table of values for one situation in exercises 7–12. Explore the following:

a. What values are possible for the first variable on this table of values?

b. What values are possible for the second variable on this table of values?

Teaching Matters: This lesson integrates geometry, data collection, and algebra, using real applications. It uses concrete materials and multiple representations including tabular, verbal, symbolic, and graphic means. Variables are introduced in contexts in which values actually vary (compared to x + 5 = 9, where students only see a fixed value for x, x = 4.) The problem situations provide an entrée into discussions of linear graphs in which intercepts and slopes have interpretations. Students are afforded natural opportunities for generating equations and solving open sentences by inspection, by tables or spreadsheets, and by working backward. In the latter instance, the steps in the process can be interpreted in terms of the situation. For example, to find the number of trapezoids for a perimeter of 305, students subtract 2 for the common edges, then divide by 3, the number of additional perimeter units added by each trapezoid. This concrete introduction to equations and their solution makes algebra more accessible to a wider range of students.

SIMULATION

Location in sample syllabus: Year 1, Unit 11

Major standards addressed: Probability, communication

Objectives: ◆ To simulate outcomes of a probability event

◆ To use simulation to estimate probabilities and expected values

Prerequisites: Simple experimental probabilities; theoretical probability of heads or tails in a coin toss

Materials: Coins

Motivating question: An obstetrician at a hospital studies recent records and finds that one week of ten births shows four girls born in a row. She becomes curious about how likely in any ten births that four (or more) in a row will be girls.

Directions for the initial activity:

Level 1

Have students work in groups of three or four, encouraging them to discuss their thinking before recording individual responses to these questions.

1. Write what you think would be a typical pattern for ten births. For example, record a boy followed by two girls by writing BGG. Continue until you have ten births recorded.

2. Repeat four more times until you have five lists of sequences for ten births each.

3. How many of your five lists have four (or more) girls in a row?

4. On the basis of your answer to exercise 3, what is your guess about the probability that a list of ten births contains four (or more) girls in a row? (If two of your five lists contained four (or more) girls in a row, your answer would be $\frac{2}{5}$.)

5. Decide how you could use a coin flip to represent a birth. (Let heads represent girls and tails, boys.) Flip a coin ten times, recording the outcomes as a list similar to that in exercise 1.

6. Repeat four more times until you have five lists of ten births based on coin flipping.

7. On the basis of your answer to exercise 6, what does the coin flipping suggest is an appropriate probability for four (or more) births in a row that are girls?

8. Is this a theoretical or an experimental probability? [Experimental]

Class discussion:

9. Pool the class results to exercise 2 to get a class guess about the probability.

10. Pool the class results to exercise 6 to get a class experimental probability for the births of four (or more) girls in a row in ten births.

11. If four girls are born in a row, we call this a *run* of four girls. Which are longer, the runs obtained from student guesses or the runs obtained from coin flipping? Which do you think is more like actual births, student guesses or modeling by coin tosses? Explain why you

Teaching Matters: **Encourage students to conjecture about this question and about how they might test their conjectures with some type of simple experiment. The majority of students will usually underestimate greatly the likelihood of long strings of occurrences in binomial outcome situations. This tendency can become part of a later discussion about the wisdom of relying on intuition in chance activities.**

think so.   [Coins are more like actual births, since the outcomes have approximately the same likelihood.]

12. Given the experimental probability determined in exercise 7 for runs of four girls, what do you think is the probability of a run of four boys (or more) in ten births?

13. The experiment using coins to represent births is called a *simulation*. Discuss how to set up a simulation to find the experimental probability that a family with four children will have three girls and one boy. Is order important?   [Using the method in exercise 5, let a family of four children be represented by four coin flips. Toss to simulate 100 families, counting how many times, t, that you get three heads and one tail. Then the experimental probability is $t/100$.]

Follow-up activities:

Level 2

Have students consult a table of random numbers. Have them work in groups and decide how they could use the data from such a table to determine an experimental probability that in twelve births, five (or more) babies in a row will be the same sex. Have each group carry out the simulation, determine an experimental probability, and then prepare a written group report. Ask groups to compare their findings and discuss any resulting differences.   [Grouping sets of numbers into lists of twelve consecutive digits can represent a birth sequence, with odd digits representing girl births and even digits representing boy births. Students should recognize that there will be discrepancies among group results and observe variation as a function of how many trials were selected in the simulation.]

Level 3

If students are conversant with a computer programming language or spreadsheet, introduce them to how the computer generates random numbers—RND(X). Using a simple program such as the following, let students explore how the experimental value of a simulation is affected by increasing the number of trials. Have them make up their own problem about births and then modify the program to carry out a simulation for ten trials. Then have students redo the simulation for 20, 30, 50, and 100 trials. Ask students to try to describe the behavior of their results as a generalization.

```
 5 REM: SIMULATION FOR X BIRTHS OF BOYS OR GIRLS
10 FOR N = 1 TO 10
20 PRINT
30 FOR X = 1 TO 10
40 LET Y = INT(2 * RND(X))
50 IF Y = 0 PRINT "G";
60 IF Y = 1 PRINT "B";
70 NEXT X
80 NEXT N
```

Level 4

Ask students to design a simulation that will determine the experimental probability that a family of five children will have all boys or all girls. Then have them find the theoretical probability by constructing a sample space, assuming that the probability of a birth being a girl is $\frac{1}{2}$. Have students contrast the two values and discuss the methods in terms of decision making.   [$\frac{1}{2^5}$, or $\frac{1}{32}$]

Level 5

Ask students to determine the theoretical probability for the motivating question.   [$\frac{256}{1024}$, or $\frac{1}{4}$]

*Teaching Matters: **All graphing calculators and some scientific calculators have a random number generator, which would be useful for this task.***

*Teaching Matters: **Students should discover that Monte Carlo experiments for simple binomial situations are fairly simple to perform and are relatively good predictors for a reasonable number of trials. They should also recognize the fallibility of their intuitions in predicting reasonable probability behaviors and the need for models and experiments to guide their decisions. The nature of this lesson makes the ideas and procedures used easily accessible to broad groups of students while providing an excellent example of how mathematics can be used to model real-world phenomena. Students working in groups, discussing their conjectures and results, and writing group reports illustrate the goals of the communication standard.***

SOLVING SYSTEMS OF EQUATIONS WITH A GRAPHING UTILITY

Location in sample syllabus: Year 1, Unit 12

Major standard addressed: Functions

Objectives: ♦ To approximate the solution of two simultaneous equations with a graphing utility

♦ To understand the connection between the graphical and algebraic representations of a function

Prerequisites: Familiarity with the zoom-in feature of a graphing utility; ability to choose appropriate windows for displaying graphs

Motivating question: Two toy rockets are launched, one ten seconds after the other. The height in feet of the first rocket after $0 \le t \le 16$ seconds is given by $h(t) = -16t^2 + 256t$. The height of the second one after $10 \le t \le 20$ seconds is given by $g(t) = -16t^2 + 480t - 3200$. How many seconds after the first rocket is launched are the rockets at the same height?

Directions for the initial activity:

Level 1

1. Use your graphing utility to plot the graph of $h(t)$ for values of t from 0 to 16. (You may wish to estimate the height of your graphing window by computing the value of $h(t)$ when h is 10, an easy value to estimate.) How high does the first rocket go?   [1024 feet]

2. Use your graphing utility to plot the graph of $g(t)$ for values of t from 10 to 20. How high does the second rocket go?   [400 feet]

3. Use your graphing utility to plot the graphs of $h(t)$ and $g(t)$ for values from 0 through 20. Without zooming, estimate how many seconds after the first rocket is launched that the two rockets are at the same height.

4. Zoom in to find when the rockets are at the same height, answering to the nearest tenth of a second. How high are the rockets when they are at the same height?   [14.3 seconds; 391.8 feet]

Class discussion:

5. Find $h(0)$, the value of $h(t)$ when $t = 0$. What does this mean in terms of the first rocket?   [It is still on the ground.]

6. Find $h(5)$ and state what this means in terms of the first rocket. Repeat for $h(16)$.   [880 feet, the height after 5 seconds]

7. Why is the formula for the first rocket only valid for values of t between 0 and 16 inclusive? (Hint: Where is the rocket at those times?)   [After 16 seconds, it hits the ground.]

8. Evaluate $h(15)$. Does this make sense? Why or why not?   [240 feet; yes, this is the height after 15 seconds.]

9. How high is the first rocket when the second one is launched? How high is the second rocket when the first one hits the ground?   [960 feet; 384 feet]

10. Find the height of each rocket 14 seconds after the first one is launched. Which rocket is higher?   [448 feet; 383 feet; rocket h]

Teaching Matters: **Working as a class, students might be asked to find the heights of each rocket for some specific values of t. Students should begin to visualize the relative positions of the rockets at some key values, recognizing why the limitations are stated for the domain values for t. Encourage students to estimate an answer to the motivating question.**

11. Find the height of each rocket 15 seconds after the first one is launched. Which is higher? [240 feet; 400 feet; rocket g]

12. What do exercises 10 and 11 tell you about when the rockets are the same height? [It occurs between 14 and 15 seconds.]

13. The rockets are at the same height (0 feet) until the first one takes off. They are at the same height (0 feet) after the last one lands. When is this? [20 seconds]

Follow-up activities:

14. Sketch each graph by hand using graph paper. Label the solution.

15. If both rockets were fired at the same time, which would hit the ground first, h or g? How many seconds earlier would it hit the ground? [Rocket g would hit 6 seconds earlier.]

Level 2

16. Use algebra to find when $h(t) = g(t)$. (Hint: What are $h(t)$ and $g(t)$ equal to? Set those expressions equal.) Does your solution agree with the results from exercises 8 and 9? Does it agree with the solution found using the graphing utility? [$3200/224$, or about 14.3 seconds]

Level 3

17. The height of a third rocket is given by $p(t) = -16t^2 + 256t - 768$. For what values of t does this equation make sense? [$4 \leq t \leq 12$] How many seconds after the first rocket is launched is the third rocket launched? [4 seconds] How many seconds after the first rocket is launched does this rocket hit the ground? [12 seconds]

Level 4

18. The general formula for the height of a rocket (t seconds after launch) under the influence of gravity is given by $f(t) = at^2 + bt + c$, where $a = -16$ (numerically half the acceleration due to gravity at 32 feet per second per second). The constants b and c can be found according to when the rocket is launched and when it hits the ground. Write a formula for a rocket that is launched four seconds after the first one and that hits the ground twelve seconds after the first one. [Hint: What is the value of $f(t)$ when the second rocket is launched? When it hits the ground?] [$p(t) = -16t^2 + 256t - 768$]

Level 5

19. The graph of the equation for the height of a rocket is a parabola. To find the maximum height of a rocket, c, and how many seconds k after launch that the maximum height occurs, we can complete the square so that the formula for the function is in the form $f(t) = a(t - k)^2 + c$. Then the vertex of the parabola is at the point (k, c). Use this method to find the coordinates of the vertices of the three parabola equations corresponding to the heights of rockets 1, 2, and 3. How do these values correspond to those from the graphing utility? (Use the TRACE option to read these values numerically from the graphs.) [(8, 1024), (15, 400), (8, 256)]

Teaching Matters: The solution to the original pair of simultaneous quadratic equations results in the linear equation $224t = 3200$, so it is not necessary to be able to solve general systems of quadratic equations to do the level 2 exercises. The level of the exercises corresponds to building a deepening insight into the significance of the functions graphed by the graphing utility. Level 1 stresses graphical understanding a bit more. Level 2, although easy, is a bit more abstract and requires algebraic manipulation. Level 3 probes understanding meaningful values of the function (i.e., the height must be nonnegative). Level 4 challenges students with the reversibility of level 3, requiring them to define the function according to when the factors equal 0—$f(t) = (t - 4)(t - 12)$. Level 5 involves the more advanced algebraic technique of completing the square.

COORDINATE TOOLS AND PROOF

Location in sample syllabus: Year 2, Unit 2

Major standard addressed: Geometry from an algebraic perspective

Objectives: ◆ To derive and apply the formulas for distance and midpoint in the coordinate plane

◆ To confirm properties of figures in the coordinate plane using numerical or variable forms

Prerequisites: The Pythagorean theorem

Materials: Graph paper

Motivating question: You are a computer game designer. In the arcade game you are designing, a player needs to toss a ball to land as close as possible to the center of a goal line. Sometimes the goal line is horizontal, but more often its location varies.

Teaching Matters: **Have students discuss this situation. They may want to look at some simpler cases (e.g., a horizontal or vertical line). They may want to try some specific cases (numerical coordinates) and generalize. Tell them that they will be able to solve this and related questions during this lesson.**

You know that the screen uses coordinates. How can you find the midpoint of any goal line? How can you determine the distance of the player's ball from the center of the goal line?

Directions for the initial activity:

Level 1

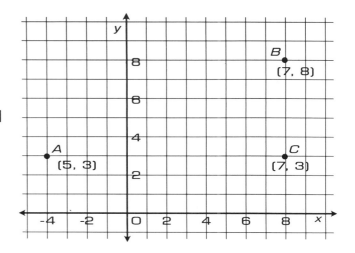

1. $A(-5, 3)$, $B(7, 8)$, and $C(7, 3)$ are points on the diagram at the right.
   a. Find the distances AC and BC. [$AC = 12$, $BC = 5$]
   b. What kind of angle is $\angle ACB$? [Right]
   c. Find the distance AB. [13]

2. $P(8, -6)$, $Q(3, -4)$, and $R(3, -6)$ are points on the diagram at the right.
   a. Find the distances PR and QR. [$PR = 5$, $QR = 2$]
   b. What kind of angle is $\angle PRQ$? [Right]
   c. Find the distance PQ. [$\sqrt{29}$]

3. In $\triangle ABC$ above, find the midpoint of each segment.
   a. Midpoint of $\overline{AC}$ [(1, 3)]
   b. Midpoint of $\overline{BC}$ [(7, 5.5)]
   c. Midpoint of $\overline{AB}$ [(1, 5.5)]

4. In $\triangle PQR$ above, find the midpoint of each segment.
   a. Midpoint of $\overline{QR}$ [(3, –5)]
   b. Midpoint of $\overline{RP}$ [(5.5, –6)]
   c. Midpoint of $\overline{QP}$ [(5.5, –5)]

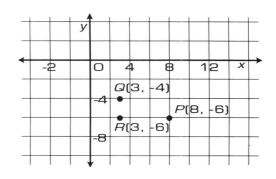

Teaching Matters: Ask students to explain their method for finding the distances in exercises 1 and 2. Discuss the need to make all values positive by using absolute value. Have them generalize verbally: when points are horizontal or vertical, the distance between them is the absolute value of the difference of the coordinates; when points are not horizontal or vertical, use the Pythagorean theorem to find the length of the side that is neither horizontal nor perpendicular.

Teaching Matters: Ask students to explain any patterns they noted in finding midpoints in exercises 3 and 4. If appropriate, make a table of their information.

Teaching Matters: This lesson blends geometry with algebra. It uses inductive reasoning as a basis for generalizing the distance and the midpoint formulas (traditionally developed for students). That this lesson occurs early in year 2 illustrates one benefit of placing the Pythagorean theorem early in the sequence. By beginning at and building on the concrete, numerical level, this lesson allows all students access to the key ideas. Teaching strategies that make ideas accessible to more students incorporate verbal and symbolic descriptions that lead to formulas without subscripts.

Class discussion:

5. Give students two points on the coordinate plane with numerical coordinates but without a triangle drawn. Have them create a right triangle to find the distance between the two points. Repeat as necessary.

Level 2

6. Give students two general points (a, b) and (c, d). Have them find the distance between them. Provide guidance as necessary. (Avoiding subscripts aids in student understanding at this stage.)

7. Ask students to write a general formula for distances in the coordinate plane. [Starting with points (a, b) and (c, d), they will find that the distance is $\sqrt{(a-c)^2 + (b-d)^2}$.]

8. Have students generalize verbally, then symbolically, the procedure for finding a midpoint.

   [x is the average of the x-coordinates: $x = \dfrac{a+c}{2}$;

   y is the average of the y-coordinates: $y = \dfrac{b+d}{2}$]

9. Have students answer the Motivating Question. [Use the midpoint and the distance formulas. If the coordinates of the endpoints are (a, b) and (c, d), the midpoint is at $((a+c)/2, (b+d)/2)$. If the ball is at (x, y), then the distance to the midpoint is

   $\sqrt{((a+c)/2 - x)^2 + ((b+d)/2 - y)^2}$.]

Follow-up activities:

Level 3

10. Have students verify for specific numerical cases that the distances from each endpoint of a segment to its midpoint are equal.

11. For various triangles with numerical coordinates, ask students to use the distance formula to classify the triangles as scalene, isosceles, or equilateral.

12. Have students use geometric coordinate software to (a) draw several quadrilaterals; (b) find the midpoints of all sides; (c) join the midpoints to create another quadrilateral; and (d) note any special properties of the new quadrilateral. [The midpoints of any quadrilateral form the vertices of a parallelogram.]

Level 4

13. Prove that your generalization for exercise 8 is correct.

TRIGONOMETRY FOR OBLIQUE TRIANGLES

Location in sample syllabus: Year 2, Unit 4

Major standard addressed: Trigonometry

Objectives:
- To recognize the correspondence of trigonometric values (sines and cosines) with coordinates on the unit circle in quadrants I and II
- To extend definitions of trigonometric functions to obtuse angles
- To recognize that supplementary angles have equal sines and opposite cosines

Prerequisites: Reflections; definitions of sine and cosine for right triangles; scientific calculator use in degree mode for finding sines, cosines, and their inverses for acute angles

Materials: Protractors, scientific calculators

Teaching Matters: **Use this situation as a problem-solving activity. Students do not have all the necessary prerequisites, but they should be able to determine that they could solve the problem if they could find the length of the hypotenuse $\overline{BD}$. Some students may suggest setting up a system of equations in which $\overline{BC}$ appears in two right triangles. Other students may recognize that it would be helpful to be able to solve oblique triangles. Let them know that some of the skills and concepts necessary for solving general triangles are the focus of this lesson.**

Motivating question: One method surveyors use to determine the height of a mountain is to measure the angle of elevation from the ground at two locations that are a known distance apart. The figure at the right supplies some data for such a situation.

What is given? What do you need to find? What are the constraints? What additional information would you like to know? Suggest sub-problems whose solution would move you closer to solving the problem.

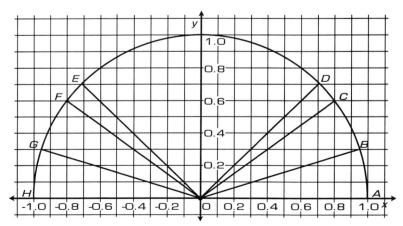

Directions for the initial activities:

Level 1

A semicircle with radius 1 and center at the origin is shown for quadrants I and II. It is half of a *unit circle.*

Complete the table below using these directions:

a. For points C, D, E, F, and G, measure the angle formed by the radius and the positive x-axis. Estimate from the diagram the coordinates (x, y) of each point.

b. Use a scientific calculator to find $\cos \theta$ and $\sin \theta$ where θ is the measure, in degrees, of the angle in part a.

c. State your observations.

Point	(a) Angle measure θ	(b) Coordinates x	y	(c) Ratios $\cos \theta$	$\sin \theta$
B	$\angle AOB = 18°$	0.95	0.3	0.9511	0.3090
C	$\angle AOC =$				
D	$\angle AOD =$				
E	$\angle AOE =$				
F	$\angle AOF =$				
G	$\angle AOG =$				

Answers:

C	37°	0.8	0.6	0.7986	0.6018
D	45°	0.7	0.7	0.7071	0.7071
E	135°	−0.7	0.7	−0.7071	0.7071
F	143°	−0.8	0.6	−0.7986	0.6018
G	162°	−0.95	0.3	−0.9511	0.3090

Class discussion:

1. Engage students in discussing whether their present definitions for trigonometric ratios can be extended to obtuse angles. Have students explore whether their calculators will accept obtuse angle inputs for sines and cosines. (This will convince students!)

2. Have students discuss their observations from the activity. The following conclusions can be drawn:

   a. The coordinates from the diagram are close estimates of the cosine (x-coordinate) and sine (y-coordinate) ratios from the calculator. With a finer grid on the diagram, there should be even closer agreement between the two sets of values.

   b. Be sure that students understand why this correspondence exists. Magnify the diagram for the first angle and complete a right triangle as shown. Note that the right triangle has a hypotenuse of 1 unit. Therefore, $\sin \theta = y/1 = y$ and $\cos \theta = x/1 = x$.

   c. Note that angles with rays in quadrant II are obtuse. Although a right triangle containing an obtuse angle cannot be formed, the unit circle permits the extension of the meaning of sine and cosine to larger angles. Radii in quadrant I reflected over the y-axis result in images in quadrant II. The angle pairs are supplementary. Students should be able to present a convincing argument why this is so.

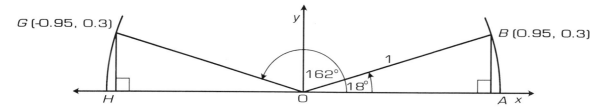

G (-0.95, 0.3) B (0.95, 0.3)

H O A x

162°   18°   1

[Reflections preserve angle measure, so m∠AOB = m∠HOG. Call this measure x. Let m∠AOG = y. Angles AOG and HOG form a linear pair, so $x + y$ = 180.]

d. When radii are reflected over the y-axis, the y-coordinates remain unchanged and the x-coordinates change signs. Thus, supplementary angles have the same sine and opposite cosines.

Follow-up activities:

Level 2

3. Without a calculator, find the sine and the cosine of the supplement of the angle given.

   a. cos 5° = 0.9962, sin 5° = 0.0872

      [cos 175° = –0.9962, sin 175° = 0.0872]

   b. cos 73° = 0.2224, sin 73° = 0.9563

      [cos 107° = –0.2224, sin 107° = 0.9563]

4. The cosine of an angle is –0.0567. About how big is the angle? [Just over 90°]

5. Find two possible angles with the given sine:

   a. sin A = 0.8746   b. sin B = 0.2588
      [61° or 119°; 15° or 165°]

6. Use your calculator to decide how to complete the following:

   a. As the size of an angle increases from 0° to 180°, its sine

      _____.

      [Increases from 0 to 1 (at 90°), then decreases back to 0 (at 180°)]

   b. As the size of an angle increases from 0° to 180°, its cosine

      _____.

      [Decreases from 1 to –1]

7. The word *cosine* comes from "complement's sine." Use five different angles to illustrate that cos A = sin (complement of A).

Level 3

8. Consider the expression (cos A)² + (sin A)².

   a. Evaluate this expression for five different angle measures.

   b. What property do your results illustrate? [The sum is always 1.]

   c. Why must this property always be true? [The cosine and the sine are two legs of a right triangle on a unit circle. This is a special case of the Pythagorean theorem.]

9. Make a table of values for the sine and the cosine of angles from 0° to 180° in increments of 15°. Use your calculator, a spreadsheet, or a computer program to generate the table.

   a. Graph the values for the sine and the cosine on two separate graphs, letting the angle value be the x-coordinate and the sine (or

Assessment Matters: Homework is a good tool for assessment. It gives teachers a chance to find out about students' thinking and to check their growth in understanding. Combined with other assessment procedures, homework can be used as a basis for adjusting instruction.

Teaching Matters: The goal of this lesson is to introduce the properties of sines and cosines of supplementary angles. This is all the information needed to solve oblique triangles (which would be the topic of the next lesson). Only the first two quadrants of the unit circle are introduced. This gives students the opportunity to recognize these supplementary relationships through numerical examples and through earlier work with transformations, without overwhelming them.

cosine) value be the y-coordinate. Connect the points with a smooth curve. (Let horizontal units represent 15° and vertical units represent 0.1.)

b. How are the graphs alike? How are they different?

c. Use a graphing utility to produce a graph of $y = \sin x$.

(1) What part of the graph on the screen is the graph you constructed?

(2) What does the graph from the utility show for negative x-values?

(3) What does the graph show for x-values over 180°?

(4) What is the general behavior of the graph?

10. Repeat part c for $y = \cos x$.

TRANSFORMATIONS WITH DATA

Location in sample syllabus: Year 2, Unit 7

Major standard addressed: Statistics

Objectives: To understand and apply data transformation properties:

- ◆ A data set with mean m and standard deviation s, translated by adding k to each value, results in a new set with mean $m + k$ and standard deviation s.

- ◆ A data set with mean m and standard deviation s, rescaled by multiplying each value by a, results in a new set with mean am and standard deviation as.

Prerequisites: ◆ Prior work with statistical measures resulting in a strong understanding of mean and range and an understanding of standard deviation as a measure of spread based on squared deviations from the mean

- ◆ Ability to determine these statistical measures with a statistical calculator, a spreadsheet, or computer software

- ◆ Understanding of the construction and interpretation of box plots

Materials: Statistical calculators or computers with spreadsheets or statistical software

Motivating question: The bookkeeper of a fast food franchise has been asked by his boss to prepare a financial profile for use in comparing her restaurant with others in the franchise. As part of the study, the salaries of all the employees have been entered into a data base and the mean and the standard deviation determined. A few days later, the boss informs the bookkeeper that all employees are receiving a 3.8 percent raise according to the new contract. How will this affect the mean and the standard deviation of the salaries?

Teaching Matters: **Students will discern that the bookkeeper could recalculate all the salaries, reenter the data, and find a new mean and standard deviation. The lesson will show students how to determine the new mean and standard deviation without translating all the data in the original set.**

Directions for the initial activity:

Level 1

A student has the following scores on nine quizzes (each quiz had a possible total of 20 points):

$$13, \quad 13, \quad 13, \quad 14, \quad 14, \quad 16, \quad 16, \quad 17, \quad 19$$

1. Enter the data values into your calculator and find the mean and the standard deviation.   [15, 2]

2. The teacher decides to adjust the scores to a basis of 25 points. One way to do this is to rescale by multiplying each score by $\frac{5}{4}$.

   a. Write the nine new scores after rescaling.

   b. Find the mean and the standard deviation of the new set of scores.

   c. How do the new measures compare with the original ones? [Each is $\frac{5}{4}$ times the original.]

3. Another way to adjust to a basis of 25 points is to translate by adding 5 points to each score.

   a. Write the nine new scores after this addition.

◆　　　◆　　　◆　　　◆　　　◆　　　◆　　　◆　　　◆

Teaching Matters: **You may wish to do a more concrete representation.**

a. Using an overhead projector, project on a centimeter grid centimeter tiles or cubes with heights 8, 6, and 10. Establish the mean [8] and show visually as illustrated that the spaces above the mean and below the mean are equal.

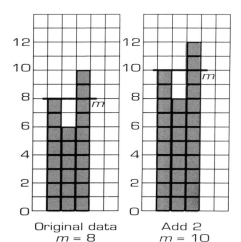

Original data　　　Add 2
$m = 8$　　　　　$m = 10$

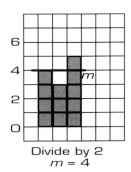

Divide by 2
$m = 4$

b. Now add two tiles to each stack. Have students decide visually that the new mean is 10 and that each stack value has been translated up by two units.

c. Return to the original tile display and divide each stack in half. Again, let students determine visually that the mean, as well as all stack values, has shrunk to $1/2$ its original value.

b. Find the mean and the standard deviation of this new set of scores.

c. How do these new measures compare with the original ones? [The new mean is the old mean plus 5; the standard deviation is unchanged.]

4. a. Construct (or have your calculator or computer draw) three box plots corresponding to these data sets: original data, rescaled data, translated data.

   b. Describe the two new plots by comparing them to the original plots in terms of transformations. [Multiplying is a stretch or scale change; adding is a slide or translation.]

   c. Are any of the plots congruent figures? Explain. [Box plots of translated data are congruent.]

   d. Are any of the plots similar figures? Explain. [Box plots of rescaled data are stretched or shrunk versions of the original box plots.]

Class discussion:

5. Discuss the results of the initial activity. Have teams of students repeat the experiment with different data sets, scale factors, and constant addends. Discuss the relative advantages and disadvantages of the two transformations.

6. Make a table to summarize the results of the various data and generalize the two properties stated in the module objectives.

7. Have students answer the motivating question. [The new mean is the original mean times 0.038. The new standard deviation is the original standard deviation times 0.038.]

Follow-up activities:

Level 2

Use data sets for which students already know the means and the standard deviations. Have them find new values under various translations and scale changes.

8. In a scale drawing of a house, 2 inches represent 1 foot. The room lengths in the scale drawing are found to have a mean of 20 inches and a standard deviation of 5.8 inches.

   a. What is the mean length of the actual rooms? [10 feet]

   b. What is the standard deviation of the lengths of the actual rooms? [34.8 inches]

9. Make a frequency distribution for the original set of data of the initial activity.

   a. Add 5 points to each score. On the same graph, but in a different color, draw the frequency distribution for the new scores. Describe the relationship between the two graphs. [The graphs are congruent; the second graph has moved up 5 units.]

   b. Multiply each score of the Initial Activity by $5/4$. In a third color, draw the frequency distribution for these new scores. Describe the relationship between the original and the third frequency distributions. [The new distribution is stretched by a factor of $5/4$.]

10. To calculate his bowling average mentally, Kyle looks at his record and subtracts 100 from each score:

Bowling scores:	125	109	112	97	127
Transformed scores:	25	9	12	–3	27

He finds the average of the transformed scores: $70/5 = 14$.

He then adds back the 100 to find his average: 114.

a. Verify that Kyle's average is 114.

b. What transformation did he use when he subtracted 100?
   [Translation by –100]

c. What property did he use in this procedure?
   [New mean = old mean – 100]

d. Find Lynette's average by using Kyle's method for her scores:

   118   127   109   121   115   [118]

11. Kyle's brother, Kurtis, used another method. First, he estimated
    that Kyle's average would be 120. Then he figured how much his
    estimate was "off" for each score: 5, –11, –8, –23, 7. He then
    averaged these "errors": $-{}^{30}\!/_{5} = -6$. Finally, he added this average to
    his estimate to get Kyle's true average: 120 + (–6) = 114.

    a. Try Kurtis's method with another estimate for the same scores.

    b. Do you think this estimation method always works? Explain.
       [Yes; it is the property of the translation.]

12. A group of students found their heights and weights:

    Height (cm)   162   155   172   163   163   166   168   170
    Weight (kg)   50   49   68   49   50   49   58   61

    a. Complete the following table:

	Mean	Standard Deviation	
Height			[165.4, 5.06]
Weight			[54.3, 6.78]

    b. Suppose the students placed 3-cm lifts in their shoes. Find the
       mean and the standard deviation of their heights.   [168.4, 5.06]

    c. Suppose each student wore 1-kg jogging weights. Find the mean
       and the standard deviation of their weights.   [56.3, 6.78]

    d. Suppose each student added 10 percent to his or her weight.
       Find the mean and the standard deviation of their weights.
       [59.7, 7.46]

    e. Complete the following table:

	Original height	Original weight	Height with lifts	Weight with jogging weights	Weight with 10 percent increase
Range					
Median					
Mode					

    f. Consider the range, the median, and the mode of a data set
       under rescaling and translation transformations of the data. What
       general statements can you make about how these statistical
       measures are changed by the transformations?   [Range: un-
       changed by a translation, stretched by scale factor in a size
       change; median and mode: translated by the amount added in
       a translation and stretched by scale factor in a size change]

Level 3

13. Consider the situation in exercise 12 again. Suppose that students increase their body weight *and* put on jogging weights. Find the new range, median, mode, mean, and standard deviation under the following conditions:

    a. They gain 10 percent in weight first and then put on 1-kg jogging weights.

    b. They put on 1-kg jogging weights and then have their total weight increased by 10 percent.

    c. Explain the difference between the results in a and b.
    [The jogging weights translate the data; the 10 percent increase scale changes the data.]

    d. Suppose that the students have a mean weight m with a standard deviation s. They increase their weight by r percent and then put on k-kg jogging weights. What is their new mean weight and the standard deviation?
    [Mean: $M(1 + 0.01r) + k$; standard deviation: $s(1 + 0.01r) + k$]

Level 4

14. The formula for the mean of a data set with n values is $\frac{1}{n}\sum_{i=1}^{n} x_i = m$.

    a. Suppose each value is increased by k. Prove that the new mean is $m + k$.

    b. Suppose each value is multiplied by a. Prove that the new mean is am.

15. The formula for the variance of a data set with n values and mean m is $\frac{1}{n}\sum_{i=1}^{n}(x_i - m)^2 = s$.

    a. Suppose each value is increased by k. Prove that the variance is unchanged under a translation.

    b. Prove that under a scale change of a the new variance is a^2 times the original variance s^2.

    c. Prove that under a scale change of a the new standard deviation is a times the original standard deviation.

Teaching Matters: This lesson connects not only statistics and algebra but also statistics and geometry through transformations. Students can see that figures, whether polygons or graphs, can be rescaled to produce figures similar to the originals or translated to produce congruent figures.

Location in sample syllabus: Year 3, Unit 7

Major standard addressed: Underpinnings of calculus

Objectives: ◆ To write a function to describe cost on the basis of given information

◆ To find the minimum of a function graphically

◆ To interpret a discontinuity of a function

◆ To realize the pitfalls of uncritical reliance on technology

Prerequisites: ◆ Using the Pythagorean theorem

◆ Understanding the minimum or maximum value of a function over an interval

◆ Using a graphing utility to graph functions and to zoom in to find an extremum of the function

◆ Understanding continuity of a function

Materials: Graphing calculators, spreadsheets (optional), programmable calculators (optional), computer with BASIC (or other) programming capability (optional)

Motivating question: A pipeline is to be built from a point R to a point S as shown on the map. Points R and S are connected by roads as shown.

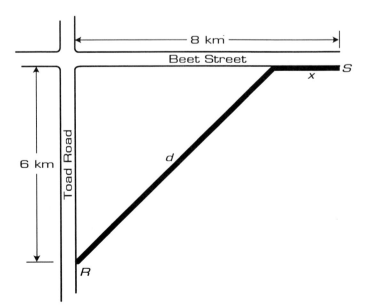

The pipeline will be built either across the land to a point x kilometers from S and then along Beet Street to S, as shown in the sketch, or entirely along Toad Road and Beet Street. If the cost is $12 000 a kilometer along the road or street and $37 000 a kilometer across the land, find x so that the cost will be minimal.

◆　　　　◆　　　　◆　　　　◆　　　　◆　　　　◆　　　　◆　　　　◆

Directions for the initial activity:

Level 1

1. How much will it cost to build—
   a. 10 km of pipeline along the road?   [$120 000]
   b. 10 km of pipeline across the land?   [$370 000]
   c. 3.5 km of pipeline along the road and 7.5 km of pipeline across the land?   [$319 500]

2. How much will it cost to build the pipeline when x is—
   a. 0 km?   [$370 000]   d. 6 km?   [$306 009]
   b. 2 km?   [$337 955]   e. 7.9 km?   [$316 831]
   c. 4 km?   [$314 811]   f. 8 km? (Hint: be careful!)   [$168 000]

3. Write a formula for the cost when the pipeline is built—
   a. entirely along Toad Road and Beet Street.   [C = $168 000]
   b. across the land to a point x kilometers from S and then along Beet Street to S.   [See class discussion.]

4. Use your graphing utility to plot the function you wrote for part 3b.
   a. What is the domain of this function? This is VERY important. [$0 \leq x < 8$]
   b. Zoom in to find the minimum cost for this function to the nearest dollar.   [$306 000]
   c. What is the minimum cost for the pipeline?   [$168 000]

5. Solve the problem as if Toad Road were not there, so that the pipeline *must* be built across land to a point x kilometers from point S. [$306 000]

Class discussion:

In exercises 1–4, what is the minimum value suggested by zooming in on the graphing utility? [$306 000] Is this the minimum cost? Explain. [No, the minimum cost of $168 000 occurs when the entire pipeline is built along the road. The graph drawn by the graphing utility, $C = 12x + 37\sqrt{36 + (8 - x)^2}$, where C is the cost in thousands of dollars, is valid only when $0 \leq x < 8$. When $x = 8$, the entire pipeline is along the road, so the cost is $168 000, less than the minimum value obtained by zooming in.]

Follow-up activities:

Level 2

Have students use a spreadsheet, a programmable calculator, or a computer program to find the cost when x = 1, 2, 3, 4, 5, 6, and 7. (Recall that the cost for x = 8 is $168 000.) They should interpret their results to find the value of x, to the nearest whole number, producing the least cost when the pipeline goes across the land and then along Beet Street. Next, direct them to use a refined search (for example, by tenths between x = 4 and x = 6) to find the value of x to the nearest tenth, which produces the least cost.

Answers:

x	1	2	3	
cost	$353 123	$337 955	$324 979	
x	4	5	6	7
cost	$314 811	$308 203	$306 009	$309 062

Teaching Matters: Bring out with students that if we rely on the graphing calculator without interpreting its information in terms of the real-world situation, we may forget that the cost function is discontinuous at $x = 8$. Consequently, we might miss the actual minimum— a mistake that in this problem would cost an extra $138 000! (In exercise 5, the graphing utility does give the answer directly.)

Assessment Matters: Today's sophisticated calculators raise many issues concerning paper-and-pencil testing. For an interesting discussion of these issues, see The Use of Calculators in the Standardized Testing of Mathematics *(1989), a book of readings edited by John Kenelly from the College Entrance Examination Board.*

Ask students to solve the problem as if Toad Road were not there, so that the pipeline *must* be built across land to a point x kilometers from point S.

Level 3

Make up a similar problem by changing the values of the distance from R to Beet Street (now 6 kilometers) and the cost a kilometer along the roads (now $12 000 a kilometer) so that it *is* cheaper to build part of the pipeline across the land. Students can use a graphing utility, a spreadsheet, a programmable calculator, or a computer program to evaluate $C = Lx + 37\sqrt{A^2 + (8 - x)^2}$ for various values of L and A. For example, when R is 5 kilometers from Beet Street and the cost along the road is $30 000 a kilometer, then $x = 1.1$ (to the nearest tenth), giving a minimum cost of about $348 000, which is cheaper than going along the roads at a cost of $390 000.

Level 4

Have advanced students apply the derivative concept in calculus to solve the original problem. (This gives rise to a cubic equation $37x^3 - 888x^2 + 8436x - 29\,588 = 0$, which is still difficult to solve analytically.)

Teaching Matters: The levels in this lesson permit a teacher to choose which approach or combination of approaches best fits a particular group of students and the available classroom resources. Regardless of the technology chosen, all students should work through exercises 1 and 2 to get a "feel" for the problem and the real-world aspects that the mathematics will model. "Grubbing with data" will enhance their consideration of constraints imposed on the mathematical model by the actual situation. This lesson carries two messages: (1) it illustrates that although technology can be helpful in solving problems, we must rely on our own thinking to apply and interpret results correctly; and (2) it illustrates how a problem may be solved in many ways by using different applications of technology.

Location in sample syllabus: Year 3, Unit 8

Major standard addressed: Discrete mathematics

Objectives: ◆ To express the terms of a series recursively

◆ To express the general term of a series

◆ To use a computer to sum a finite series (and thus estimate the sum of an infinite series)

◆ To investigate numerically using a computer whether a series converges or diverges

◆ To compare the sum of a series with known values

Prerequisites: ◆ Experience using spreadsheets, including the use of simple formulas

◆ Familiarity with e and $y = \ln x$

◆ Ability to evaluate $n!$, 2^n, and $(-1)^n$

Materials: Computer spreadsheet, computer programming capability

Motivating question: According to one version of Zeno's paradox, if you are in your classroom and try to leave, you can never get out the door! If you start walking toward the door, before you can get out, you must go half way. Once there, before you can get out, you must go half the remaining distance to the door, and so on. You will go $\frac{1}{2} + \frac{1}{4} + \frac{1}{8} + \frac{1}{16} + \ldots$ of the distance. Complete the following chart:

Term number	1	2	3	4	5	6	7	8
Value of term	$\frac{1}{2}$	$\frac{1}{4}$	$\frac{1}{8}$	$\frac{1}{16}$				
Sum	$\frac{1}{2}$	$\frac{3}{4}$	$\frac{7}{8}$					

Describe what is happening to the sum. Can you get out the door? Explain!

Directions for the initial activity:

Level 1

1. Set up a spreadsheet like this:

	A	B	C
1	Zeno's Sum		
2	term no.	value of term	sum
3	1	0.5	0.5
4	2	0.25	0.75
5	3	0.125	0.875
6	4	0.0625	0.9375
7	5	0.03125	0.96875
8	6	0.015625	0.984375
9	7	0.0078125	0.9921875
10	8	0.00390625	0.99609375
11	9	0.001953125	0.998046875
12	10	0.0009765625	0.9990234375

Spreadsheet showing values

◆　　　◆　　　◆　　　◆　　　◆　　　◆　　　◆　　　◆

	A	B	C
1	Zeno's Sum		
2	term no.	value of term	sum
3	1	.5	.5
4	=A3+1	=B3/2	=C3+B4
5	=A4+1	=B4/2	=C4+B5
6	=A5+1	=B5/2	=C5+B6
7	=A6+1	=B6/2	=C6+B7
8	=A7+1	=B7/2	=C7+B8
9	=A8+1	=B8/2	=C8+B9
10	=A9+1	=B9/2	=C9+B10
11	=A10+1	=B10/2	=C10+B11
12	=A11+1	=B11/2	=C11+B12

Spreadsheet showing formulas

From the spreadsheet formulas, we can see that the values of row 3 of the spreadsheet are entered as numbers and that the remaining rows are formulas. In column A the *term number* is always 1 more than the number in the cell immediately above it. In column B the *value of the term* is half the value in the cell immediately above it. In column C the value of the *sum* is the value of the sum in the cell immediately above it plus the value of the new term that is added to get the new sum. To make this spreadsheet you need only type in rows 3 and 4 and then fill down (automatically) from row 4 as far as you wish to go. Complete the spreadsheet to find the sum of the first twenty terms of the series. As the number of terms increases, the sum approaches what number? [1]

2. Make a spreadsheet as in exercise 1 for $1 - \frac{1}{2} + \frac{1}{4} - \frac{1}{8} + \frac{1}{16} - \frac{1}{32} + \ldots$. As the number of terms increases, the sum approaches what common fraction? [$\frac{2}{3}$]

3. Make a spreadsheet as in exercise 1 for the *harmonic series*

$$1 + \frac{1}{2} + \frac{1}{3} + \frac{1}{4} + \frac{1}{5} + \frac{1}{6} + \ldots.$$

[As the number of terms increases, it is not clear that the sum approaches a limit.]

This is a famous example of a series that *diverges* (has no finite sum) even though the succeeding terms get very small. How many terms are needed before the sum exceeds 2? Exceeds 3? Exceeds 4? [4; 11; 31]

Each of the following series *converges* (has a finite sum). Use a spreadsheet to find the sum of the first twenty-five terms and then guess the exact sum. (Hint: Choose answers from e, $1/e$, $\pi/4$, and $\ln 2$.)

4. $1 - \frac{1}{2} + \frac{1}{3} - \frac{1}{4} + \frac{1}{5} - \frac{1}{6} + \ldots$ [ln 2]

   (Hint: You may want to include a separate column on your spreadsheet that alternates 1, –1, 1, –1, 1, –1,....)

5. $1 - \frac{1}{3} + \frac{1}{5} - \frac{1}{7} + \frac{1}{9} - \ldots$ [$\pi/4$]

   (Hint: You may want to include a separate column on your spreadsheet for the denominators.)

6. $1 + \frac{1}{1!} + \frac{1}{2!} + \frac{1}{3!} + \frac{1}{4!} + \frac{1}{5!} + \frac{1}{6!} + \ldots$ [e]

7. $1 - \frac{1}{1!} + \frac{1}{2!} - \frac{1}{3!} + \frac{1}{4!} - \frac{1}{5!} + \frac{1}{6!} - \ldots$ [$1/e$]

Teaching Matters: Encourage students to discuss their discoveries in exercises 1–7. As they state conjectures, ask questions to help them formulate their thoughts into testable hypotheses. Direct their thinking about series with questions such as the following: If the terms of a series get as small as we like, does this necessarily mean that the series will converge? [No, consider the harmonic series.] Suppose that the terms of a series do not get as small as we like. For example, suppose the terms never get smaller than 0.001. Can such a series converge? [No] Discuss the sum of $1 - 1 + 1 - 1 + 1 - 1 + \ldots$, by viewing it first as $(1 - 1) + (1 - 1) + (1 - 1) + \ldots$ and then as $1 + (-1 + 1) + (-1 + 1) + (-1 + 1) + \ldots$.

Follow-up activities:

Level 2

A spreadsheet can be used to determine the sum of a series by using a general term for the series. For example, a general term for the series in exercise 1 can be expressed as $\dfrac{1}{2^n}$. Redo exercises 1–7 by using a general term for each series as a formula on the spreadsheet. Then find the sums of the series below in a similar manner.

8. $\dfrac{1}{1\cdot 2} + \dfrac{1}{2\cdot 3} + \dfrac{1}{3\cdot 4} + \dfrac{1}{4\cdot 5} + \ldots$ [1]

9. $\dfrac{1}{1\cdot 3} + \dfrac{1}{3\cdot 5} + \dfrac{1}{5\cdot 7} + \dfrac{1}{7\cdot 9} + \ldots$ [½]

10. $\dfrac{1}{1\cdot 3} + \dfrac{1}{2\cdot 4} + \dfrac{1}{3\cdot 5} + \dfrac{1}{4\cdot 6} + \ldots$ [¾]

11. $1 + \dfrac{1}{2} - \dfrac{1\cdot 1}{2\cdot 4} + \dfrac{1\cdot 1\cdot 3}{2\cdot 4\cdot 6} - \dfrac{1\cdot 1\cdot 3\cdot 5}{2\cdot 4\cdot 6\cdot 8} + \ldots$ [$\sqrt{2}$]

Level 3

Have students explore series by using a computer language to apply the general terms of series in finding their sums. Sums can be found by initializing (e.g., set SUM = 0) and incrementing (e.g., replacing SUM by SUM + $\frac{1}{N}$) within a loop. Products can be found similarly. For example, to compute $n!$, initialize PRODUCT to 1 and replace PRODUCT by PRODUCT * N. Have students redo exercises 8–11 by writing programs that sum the series for values of the loop index N. Encourage them to discuss the intuitive notion of a limit as larger index numbers N permit the computer to sum more terms of the series.

Level 4

Have students compare the harmonic series to the following series to show that this series diverges:

$$1 + ½ + (¼ + ¼) + (⅛ + ⅛ + ⅛ + ⅛) + \ldots$$

FINAL CHAPTER NOTE

In chapters 3 and 4, we have tried to offer diverse examples of how a core curriculum could be implemented in grades 9–12. Though the approaches vary, there is one commonality. Each uses a developmental approach to focus on concept development rather than on by-hand calculation and symbol manipulation. We are convinced that this notion is central to teaching a common core of mathematics to all students. The curriculum models and abbreviated lessons are intended to be illustrative rather than prescriptive. We hope that these examples will stimulate your thinking to find even more creative ways of bringing the richness of a core preparation to everyone.

The following chapter looks at many issues that must be resolved in introducing and managing a core program in mathematics. It looks at people factors, planning issues, and the management of the change process.

Teaching Matters: This lesson makes full use of technology to allow students to investigate mathematical ideas. The application of a spreadsheet emphasizes the relationship between successive terms of a series and the notion of partial sums in the way that it is constructed. The output in the form of numerical data gives students concrete information on which to understand the underlying concepts. Using a general term for expressing series' terms on the spreadsheet and within computer programs provides an excellent introduction for later work involving sigma notation. This work can be related back to how series terms are generated from the general term and the index values. The lesson further serves as a precursor to many ideas studied in calculus, such as the notion of limits and the comparison test for series.

CHAPTER 5
CHANGING TO A CORE CURRICULUM

Innovative schools of the past have developed curricula to respond to the needs of the day. None have offered a curriculum that anticipated the world of tomorrow.
—Harold G. Shane

A *core curriculum* is a significant redirection of the focus of high school mathematics programs. It is a professional commitment that *all students*—regardless of their background, motivation, or future pursuits—need to achieve a dynamic form of mathematical literacy rooted in an acceptable proficiency with a common set of learning objectives. Although the means of achieving these objectives may be negotiable (curriculum design), the outcomes associated with those objectives are not. Achieving a balance between these two curricular foci is the challenge that must be met locally to accomplish mathematics education reform. The purpose of this chapter is to offer guideposts that may smooth the transition from current programs to a core curriculum.

IMPLEMENTING A PLAN FOR CHANGE

Those few districts that have made progress toward a core curriculum at grades 9–11 have done so by virtue of sustained collegial efforts. Careful planning, communication, consensus building, and problem solving along the way have been common key steps in their success. The following set of guiding points are divided into six phases for implementing a core curriculum: *awareness, commitment, planning, development, implementation,* and *evaluation.* Two phases—planning and development—should proceed at the same time, whereas the others may significantly overlap. The guiding points in the following sections are not intended to be exhaustive. We hope they will stimulate the thinking of a leadership group in developing a local plan of action for a core curriculum.

Awareness

The goal of the awareness phase is to raise the consciousness of the people instrumental to effecting change. These people must be convinced of the need to implement a core curriculum.

♦ Use department meetings to build a common awareness of the need for major program changes.

- Study a few significant reports (e.g., NAEP data) and publications (e.g., *Everybody Counts* [Mathematical Sciences Education Board 1989]) as a department with one or two people leading a discussion on the implications for your school and district.

- Over several department meetings, have staff members volunteer to present two 9–12 standards from the *Curriculum and Evaluation Standards for School Mathematics* (NCTM 1989) and illustrate with activities from their own teaching the changes intended in instruction.

- Invite a local employer in to address the department members and key administrators about the realities of current job tasks and the preparation of students entering the work force.

♦ Encourage staff members to take an in-depth look at the existing program in terms of changing expectations.

- Consider how the current curriculum compares to the *Curriculum and Evaluation Standards* and other reports. Make a list of changes that are needed. Divide it into "things we can affect" and "things we can't affect." Begin to work with the first list, but keep the second list handy for later attention (perspectives change and so do barriers).

- Examine existing data to determine what happens to students; re-examine course sequences to decide their curricular rationale and what they accomplish.
- Gather input about the appropriateness of the program for meeting the range of diverse needs of students. Find out about employment test questions, on-the-job training, mathematics-related skills applied in local businesses, and additional training that graduates have to obtain in mathematics.

◆ Use this introspective period to bring to the surface other issues pertaining to the program.

- How do changes in the curriculum take place? Are they added on? Are instructional problems "solved" by restructuring curriculum? Does the program target exit outcomes or, more ambiguously, "accomplish the most that it can"?
- Do course staffing practices and attitudes toward student subpopulations reflect a comprehensive program for *all* students? Are beliefs about students consistent with research findings and calls for change?
- How does the department work together? What are meetings used for? Is the department a confederacy of individual interests or do staff members work and plan together? Who monitors program results in terms of data? How is the data used in decision making?
- How does the need to implement a core curriculum interface with other pressures the program may be facing?

Commitment

The goal of the commitment phase is to translate an awareness of and a concern for change into an active group of persons committed to working over an extended time toward reforming curriculum and instruction.

◆ Expand the reform working group to include the representation required to accomplish major change.

- Develop a power base of support for change. The magnitude of curriculum restructuring precludes working solely within the department. Administrative, community, and recognized expert resource persons will be necessary to mobilize and continue to drive the change over an extended time period.
- Elicit input from persons on the periphery of the program as to its effectiveness and its weaknesses. Gather information from guidance personnel, feeder schools, vocational schools, and other high school departments that draw on mathematics skills.
- Explore resources and sources for support from the district and community levels. How much support are they likely to give? In what form will it be? Are there funds (or expertise) available from business-education partnerships? Who are the key people representing parent groups that can be supportive or a problem?
- Are there groups representing special populations of students (e.g., special education) who may have input and may need to become a continuing part of communications?

◆ Initiate the goal-setting process, expecting revisions to occur as time and events proceed.

- Goals drive the planning for change. Long-range goals have to be visionary; they must be set high enough and idealistically enough to capture support and point the direction for change for several years.

The key is the commitment to begin with as bold a vision, and as powerful a statement to support that vision, as possible—no more general math.

- Short-range goals have to be manageable, balancing the need for accomplishment with the frustrations of change.
- Circulate the broad goals widely and solicit input and suggestions from many audiences. Support and understanding are garnered through involvement.

♦ Continue explorations that will be helpful in instituting a core curriculum.

- Learn what you can from districts that have accomplished major curriculum reform about the process used. Look to other districts not for programs but for suggestions on how they brought about the changes that resulted in their programs.
- Have someone from outside the department discuss the process of change and how humans react to change—its characteristics, phases, pitfalls, and advantages. Consider the implications for the department.
- Send department members to professional meetings where they can learn more about teaching a core curriculum to expanded populations. Seek out similar information from professional journals and other publications.

Planning

The planning phase will be ongoing with the development phase. Both phases will continue through full implementation. The planning activities begun initially will translate the long-range goals into a structure to oversee long-term implementation, with details handled as work proceeds.

♦ Establish a leadership team to monitor the implementation of the *Curriculum and Evaluation Standards* within a core curriculum.

- Most of the planning and execution of a core curriculum will be done by department members. The leadership team is needed to provide longer-range input and management, to address implementation issues, and to tackle ongoing and summative evaluation.
- The team needs to develop a working framework that can give continuing direction and support to developers and implementers. Decisions can be made by different groups, but they should fit within a review process that keeps the curriculum revision on target.
- A time line for major activities should be part of this framework. The time line will furnish the general plan of action to direct activities.

♦ Identify the resources and obstacles that relate to achieving the goals.

- Examine the constraints within which change will have to *begin*. All problems do not have to have a solution before you begin to work on them. Constraints have a way of changing with events and success— don't compromise your goals because there is *presently* no apparent way to overcome certain problems.
- Identify the people who will perform various roles in the implementation, such as getting policies changed, trying out new ideas and getting them to work, and managing changes administratively. How can everyone be involved in ways that will make them supportive?
- Make a list of resources needed at different points on the time line— personnel, financial, other. Which items are tied to funding? Where can additional support be found? What kinds of assistance from outside the district are available?

♦ Anticipate the effect that changes will have on people and procedures.

- Identify "attitudinal" changes that will have to accompany curricular and instructional practices. Do you need the credibility of persons from outside the district, particularly the business world, to convince some individuals to buy in?

- Examine how the roles that people hold in the department will change. How can feelings about loss of ownership or status be addressed? How can assistance and leadership be provided in a non-intrusive and nonmanipulative way?

- List the programmatic requirements and descriptions that must be changed, such as requirements for graduation, curriculum descriptions used for guidance, and testing procedures.

- What factors influence how the program is judged, such as achievement (e.g., SAT, ACT, state) tests? How will these be impacted by the initiation of a core curriculum? How can transition problems be anticipated?

- Consider how a core curriculum impacts the articulation of programs: students entering the district, coming from feeder schools, going on to vocational schools, exiting to post-high school programs.

- Consider how the nonacademic community (particularly parents) may react to changes. Newsletters, parent nights, or newspaper articles can head off nostalgic resistance to change and build support among those who want educational reform.

♦ Plan a support structure for assisting students and staff members who encounter difficulties.

- Some students will be at risk in a core curriculum, particularly during transition. What kinds of programmatic provision (e.g., extra assistance in a mathematics lab) can be provided? Consider other support means that give individual reinforcement: personal advocates, nonschool mentors, parental involvement, community support.

- Provisions have to be made for learners with behavioral or learning handicaps whose needs cannot be met in the core curriculum. How will they be identified? What provisions will be made? How can these provisions be made without compromising the integrity of the core curriculum?

- Plan how to support staff members who experience anxiety or difficulties in adjusting. Consider teaming, individual pacing of changes, department meetings with special focuses (e.g., how we feel about changes), and individual growth and support plans.

Implementation

Planning and implementation will overlap considerably. Rather than being considered as separate events, they might be viewed as a continuous cycle governing the process of change: *planning, implementation, assessment.*

♦ Phase in changes over a transition period.

- Identify a time line with specific focuses for implementing a core curriculum. For example, Year 1: ninth-grade component, with experimentation at tenth- and eleventh-grade levels; Year 2: tenth-grade component, with ninth-grade improvement and upgrading consistent with changes in grades 5–8; Year 3: eleventh-grade component, with upgrading in tenth grade and the beginning of summative assessment in ninth grade.

- Purchase the materials to implement and support the program over a phase-in period. This will have to be anticipated in the planning period as a departure from traditional instructional purchase plans.

- Monitor the reactions of everyone involved in the changes to furnish necessary support and to use this information in the next phases of the implementation.

- Minimize factors that might inhibit changes.
  - Postpone the central collection of summative data that compare groups or individuals. People will be anxious about performances; measures of achievement may not quickly reflect the power of changes; the means of measuring achievement may not be appropriate for the revised outcomes expected.
  - Make others aware that some measures of program success can easily be misinterpreted if viewed too narrowly (e.g., there is usually an inverse correlation between SAT or ACT scores and the number of students tested).
  - To the extent possible, pilot changes with those staff members who are most dedicated to change, but involve all department members in the planning and sharing of results. Then use pilot teachers to assist other staff members as the programs become institutionalized throughout the school.
  - Schedule periodic discussions among staff about the effects of change on individuals. Areas of concern might include the need for stability, a frequent longing to go back, uneasy feelings of doing things for the first time, extra expenditures of time and energy, and the difficulty in changing expectations in students and ourselves.
- Maintain communication as a top priority.
  - Keep all parties informed of the progress of the changes—nonpiloting staff, parents, nonmathematics faculty, administrators, students, and the general public. Encourage the frequent involvement of policy-implementing administrators in group meetings to show their support and interest in communication.
  - Have staff members report to the department the successes and discoveries of new programs as well as the problems requiring their professional input and expertise.
  - Maintain a file of all proceedings of the various planning, implementation, and evaluation groups and make it readily accessible to anyone wishing to know exactly how decisions are being made and how the work of these groups might affect individuals.
  - Establish various mechanisms that encourage individuals to get their questions or concerns readily addressed or to air in nonthreatening ways their feelings about changes.
  - Network with other school systems that are implementing a core curriculum.
- Apply supportive structures to help individuals with implementation.
  - Work as teams to develop new materials, instruction, or evaluation means, professionally supporting one another and sharing the successes or problems.
  - Address needs as a department in developing or seeking expertise needed for implementation. For example, individual staff members can be sent to meetings or to other districts to acquire information or expertise that can be later shared with the whole staff.
  - Plan changes in the program that consider each staff member's needs and pacing. For example, a five-year plan for implementation might contain a consolidation year in which additional changes are curtailed while staff members "get their feet under them" in accommodating changes already underway.
  - Assist individuals who wish to "get up to speed" through more formal avenues such as courses or professional staff development experiences.

- Assess holistically the impact of program implementation.
  - Collect and share information about the effects of program changes on such areas as the number of students being instructed in new topics, electing additional coursework, and raising expectations about their future careers.
  - Gather and share input about program successes from students, parents, guidance personnel, community representatives, and other faculty members.
  - Reflect as a staff the effect of changes on professional views about students, instruction, curriculum, department relationships, work within the school structure, and evaluation.
  - Plan regular means to get information back to planning groups about the effects of implementation and the problems that need resolution.

Evaluation

Evaluation involves assessing the ongoing and summative effects of program changes and student achievement as well as the changes in the ways that students are evaluated.

- Establish criteria to measure the ongoing effectiveness of the core program.
  - Compile data on the quantity and quality of mathematics being taken by students and the level of preparation achieved, according to students' future interests.
  - Assess changes in students' attitudes toward mathematics, work habits and study behaviors, and responsibility for learning.
  - Identify the students experiencing limited success. What provisions are made to intervene with additional assistance? When does it occur? What adjustments are necessary instructionally or programmatically to increase their success? What is the dropout rate?
  - Gather data from outside the department on program results—from employers, colleges, other departments, parents, and students—on program results.

- Consider the effect of a core curriculum on instruction and professional growth.
  - Identify how teaching techniques have changed and how attitudes have been modified toward instruction and toward students. For example, how much class time is used for instruction?
  - Identify any changes that have affected the way the staff functions: cooperative teaching or testing, the character of department meetings, attitudes about the program, approaches to solving instructional problems, discussions about teaching or mathematics.
  - What kinds of professional growth activities are teachers engaged in? How is the staff adjusting to change, collectively and individually?
  - Have any changes occurred in staff perceptions about their roles in relation to program goals? For example, is there more awareness of responsibility for program goals as well as accountability by teaching assignment?

- Realign evaluation methods with program goals and instructional techniques.
  - Review how evaluation data is used to make decisions about student learning and intervention activities. Assure that evaluation is used to make appropriate decisions about how students learn as well as to provide a measure of their progress.

A Michigan State University study of general mathematics classes showed that only ten minutes of each class period were used for instruction whereas most of the period was used in ninth- and tenth-grade college preparatory classes.

- Target evaluation as a major thrust of staff development activities in incorporating the *Curriculum and Evaluation Standards* and a core curriculum. Disseminate the changes in evaluation approaches and goals widely among interested audiences beyond the instructional staff.
- Revamp evaluation means at the classroom level and the program level to assure that evaluation matches the revisions of the core curriculum by incorporating technology; by emphasizing concepts, problem solving, communication, and connections; by including applications and performance tasks; and by using multiple methods of evaluation of individual work and group work.
- Compile achievement data on students from multiple sources in gauging the impact of a core curriculum. Actively help administrators and others interpret the results accurately.

CRITICAL ISSUES

Certain issues loom large on the landscape of factors likely to influence successful implementation of a core curriculum. For purposes of discussion, these issues have been grouped into these clusters: *Transition, Phasing In, Administrative Size, Assessment, Opportunity to Learn,* and *Leadership for Change.*

Transition

The core curriculum assumes a preparation consistent with the K–8 standards. Many students in the next few years will arrive at grade 9 with an understanding of mathematics that will fall short of this goal. The dilemma is how to manage the curriculum for these students during the transition. The most obvious alternative is to phase in the core curriculum gradually, filling in the gaps in the students' preparation in order that they might achieve *some* core outcomes.

Although this seems to be a practical solution to the transition problem, it is also fraught with risks unless it is truly recognized as a short-term transition device. The danger here is that the curriculum becomes the variable and the students' exit skills become the compromised result. Unless the focus is kept on achieving core outcomes, independently of time requirements, the danger is that a core curriculum will be subject to the same erosion of accountability as a traditional program. It is inconsistent with the core principle to accept this alternative as a long-term solution.

A better transition strategy is to make time the variable, with the core the target for all students. We have inherited a system that makes the amount of time spent in the program the requirement for graduation, but this need not be so. Increasingly, states and communities are looking at exit competencies as a measure in certifying the credibility of students' education. The three-year core curriculum is the curriculum means to assure that students can demonstrate their competence on such exit assessments. It therefore should become the minimum graduation requirement, with entry time into the program varying according to student readiness. The following multientry approach to achieving the core program will ease the transition until the K–8 standards are in place.

In a multientry strategy, students begin the core when they are ready for it. Graduation requirements depend on completing the core. Time is the variable for achieving the core. A transition course to bridge students' preparation from a traditional elementary school program to a core program can be an enabling means for multientry. A transition course

The historical fact that graduation requirements are stated in terms of Carnegie units of mathematics, rather than in terms of competencies, has placed the responsibility for accomplishment primarily on the curriculum rather than on the student.

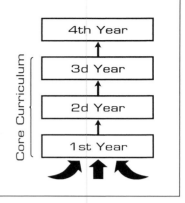

Students enter the core curriculum from different grade levels (year in school) but with equivalent preparation. The strategy assumes either a sound prerequisite preparation consistent with the K–8 standards or a transition course before entry into the first year of the core. Students may begin the first core year after grade 6, 7, 8, or 9, depending on the strength of their preparation and the depth of their understanding.

Fig. 5.1. Multientry core strategy. The core curriculum comprises three years of mathematical study. Entry time into the first year is the variable.

must change students' orientation from a skill-oriented, teacher-dominated elementary school approach to the problem-solving, student-centered, concept-oriented approach characterizing the core. In this transition course, students would learn to take more responsibility for their own learning while having communication skills and group activities stressed. Working with models and applications would deepen their understanding of concepts and connecting ideas. Ample experience with technology would prepare them for its increased application in the core program.

Support Systems

The core curriculum extends topics and expanded outcomes to much more diverse student populations than have been served in the past. Many students may be at risk for achieving success in this curriculum unless means in addition to primary instruction are found to support them. The developmental learning approaches illustrated by lessons in chapters 3 and 4 suggest how it is possible to design learning activities for core topics for all students. But they do not reflect the kinds of additional support that many students may need to achieve those outcomes. The diversity, particularly during transition, of students' attitudes, backgrounds, and understanding, along with external factors influencing students' learning, makes it imperative to rethink how we accommodate students' needs in the learning process.

A richer variety of safety nets need to be available to students at the time that they begin experiencing learning difficulty in the core. We must accept the premise that learning difficulties are best addressed through instructional means and support systems rather than through lowered expectations or downgraded curricula. We must also be realistic about our expectations of meeting students' individual learning needs within the framework of classroom teachers' primary responsibilities. Organizationally, the school needs to consider how to expand its services for students at risk instructionally. Teachers may need to become more aggressive in the early diagnosis of learning difficulties and the referral of students for additional learning support.

Some schools have found success in forming study groups or in referring students to mathematics laboratories, staffed by teachers and upper-level students, for additional assistance prescribed by the classroom teacher. Students needing extra instructional time, individual attention or prodding, or alternative presentations of ideas get that help when they need it. At-risk students may also have a designated faculty advocate or community mentor who regularly interacts with them and who helps them resolve problems. Another alternative is intervention assistance teams, small groups of faculty who meet regularly to discuss and find additional support for the at-risk students under their guidance. Programs to increase the involvement and support of significant nonschool adults for at-risk students may be another means for meeting needs, such as socioeconomic or cultural differences among the student population, within a community.

Phasing In

Instituting any core program merits a phase-in plan to assure a smooth transition from the existing curriculum. A plan and a schedule for this transition need to be adopted and supported by whatever policy changes are required to implement them. For example, a school or district may need to purchase instructional materials over several years instead of purchasing all mathematics materials in one year of a cycle. Staff development needs will be extensive, requiring planning, funding, and, possibly incorporation into negotiated agreements. Requirements affecting graduation and state and other local requirements (e.g., vocational schools) will have to be phased in. Changes in local testing programs should follow an appropriate time schedule coordinated with the phase-in plan. The following five-year schedule is one example of a transition plan.

Example of a Five-Year Transition Schedule

Year	Primary Activity	Secondary Activity	Related Activities
0	Develop an awareness of the need for a core curriculum.	Set up coordinating and planning groups.	Explore multiple sources for ideas, materials, and pedagogy.
1	Revise curriculum guides; develop a transition course as necessary.	Introduce some core units into all courses; conduct planning and coordinating groups.	Review policy implications; do staff training; reformulate graduation requirements.
2	Adopt materials for grade 9.	Implement teaching changes in existing grade 9 courses.	Provide extensive staff training and planning; raise public awareness.
3	Implement the grade 9 core; adopt materials for grade 10.	Implement teaching changes in existing grade 10 courses.	Begin formative evaluation; train staff in evaluation; make first core year a graduation requirement for ninth graders.
4	Implement the grade 10 core; adopt materials for grade 11.	Implement teaching changes in existing grade 11 courses; revise and refine grade 9.	Monitor staff reactions to changes; prepare summative data on changes (nonachievement data for grade 9); make two core years a graduation requirement for ninth graders.
5	Implement the grade 11 core.	Revise and refine grade 10; prepare summative achievement data on grade 9.	Prepare summative data on changes (nonachievement data for grade 10); make three core years a graduation requirement for ninth graders.

The schedule above provides two years for phasing units of the core into current courses. In years 3 to 5, students become accountable for achieving part or all of the core as a graduation requirement.

Administrative Size

In this document, considerations for managing change have been directed generally at schools with a mathematics department of four to ten members and staffs who have autonomy for restructuring their curriculum for the population of that building. Schools within a larger district with a central organization and schools with one to three staff members face additional challenges in achieving a core curriculum.

Large Districts

Districts with multiple high schools typically coordinate curricular changes from a central office, with program variations determined largely by student enrollments. Top-down restructuring, though efficient for management and economy reasons, is not particularly conducive for changing professional attitudes, managing personal effects of change, or building the program ownership necessary for finding solutions to problems. Large urban districts also are likely to have more students potentially at risk as program requirements are raised. Problems in developing appropriate support for student success may be more pronounced, more widespread, and more site-specific in large districts.

We believe that one of the most significant components in changing to a core curriculum is the professional growth of teachers as they participate in the awareness, planning, and implementation phases. These are dimensions that leadership in large districts are well advised to build into their planning for a core curriculum. A combination of coordinated change implemented through site-based management may satisfy these two seemingly disparate approaches. A central office can most efficiently coordinate committees for course-of-study revisions, develop support materials, conduct staff development training, order materials, and develop instruments or guidelines for managing various phases of changing to a core curriculum. But individual school personnel need to have considerable latitude in controlling the schedule of changes, the type of staff development needed, and the manner in which a core curriculum is phased in.

Small Staffs

The difficulty for schools with one to three mathematics department members is finding enough people to implement a curriculum change of this magnitude. At the same time, these teachers need the opportunity for professional dialogue and collegial support to manage change. The challenge is to find ways to work together across districts to satisfy these needs.

Sometimes small schools are part of larger organizational units such as county districts or regional service centers. These larger units furnish certain services and usually assume a coordinating responsibility for revising courses of study, organizing staff development opportunities, and facilitating review of instructional materials for adoption. With this kind of assistance, teachers at small schools should seek to get these services coordinated and extended over a multiyear period. Larger-unit personnel would be expected to coordinate the phases required for implementation, regularly bringing teachers together to find group solutions to implementation problems. To meet the need for more daily communication and sharing of ideas, teachers might look into the possibilities of networking through avenues such as PSInet or other available communication means.

Where small schools are not part of a larger unit, individuals may need to take the initiative to bring this problem to a local institution of higher learning to explore the possibilities of linking university coursework to the curriculum development and implementation needs of the local schools. Many colleges of education and university mathematics departments provide service courses specifically tailored to assist individuals and districts with emerging educational needs. A sequence of several courses might be designed around developing and implementing a core curriculum. Participants get college credit for work done to create the products and acquire the expertise required to reform instruction in their local schools. An added advantage is that people with expertise in curriculum development and instructional delivery are available on a continuing basis to help solve problems and to react to proposed approaches.

Other sources of assistance are state and provincial educational agencies (SEAs) and funding sources focusing on mathematics improvement. SEAs have the responsibility of assuring quality educational opportunities for students, usually through local schools and support of education. But in extenuating circumstances, more direct help in addressing specific problems can be requested from SEAs. That help might come in the form of directed state or provincial aid to meet the problems of smaller districts. Development projects and teacher enhancement experiences to meet a state- or province-wide problem are other alternatives. Inquiries might also go to an SEA for help in pooling financial resources for consortium approaches to implementing a core curriculum or to solicit discretionary monies (e.g., Eisenhower funds) from the SEA for addressing a need through designated projects.

Assessment

Countries in North America stand alone among the nations of the world in the reliance placed on assessing students' understanding through closed-form (multiple-choice) instruments. Customary methods for measuring the achievement of students and making decisions about the effectiveness of programs have grown out of nineteenth-century psychology and the historical practice of using multiple-choice tests to categorize large groups of people.

A significant challenge in implementing a core curriculum is to begin to bring assessment methods and tasks into alignment with the learning outcomes intended by the *Curriculum and Evaluation Standards* and with what we now know about how students learn mathematics. Standards 1–4 *(Problem Solving, Communication, Reasoning, Connections)* specify learning goals that aim for a deeper understanding of mathematics than just discrete manipulative skills. Accurately assessing students' learning relative to these and other goals of the *Curriculum and Evaluation Standards* will require substantially different classroom techniques and program monitoring than we have used with past programs. Performance tasks, open-ended questions, student portfolios, holistic scoring, investigations, interviews, and student projects are just a few of the many possible alternatives being incorporated into mathematics assessment.

The ways that students study and use mathematics must also come to be a part of realistic assessment practices. Asking students to respond to assessment tasks without recourse to the technological or developmental tools used in their study of ideas is unrealistic, unsound, and a poor measure of students' capability. Failing to assess how students perform in group projects and problem-solving tasks is failing to acknowledge how they will apply mathematics in the real world. Expanded assessment alternatives will be very much in the interest of student populations perceived to be at risk. Alternative assessment methods will enable students to show what they know rather than what they don't know, giving teachers more accurate feedback in planning instruction and making evaluations.

It is only a slight exaggeration to describe the test theory that dominates educational measurement as the application of twentieth-century statistics to nineteenth-century psychology.

—R. J. Mislevy

Assessment Matters: Remember to assess affective as well as cognitive outcomes. The attitudes and beliefs of students about themselves, about schooling in general, and about mathematics can be important factors in whether and how students learn mathematics. Knowledge about these attitudes and beliefs, in turn, suggests how a teacher should organize instruction to obtain desired outcomes.

Fortunately, major entities assessing student performance (e.g., ETS, NAEP) are beginning to explore seriously the kinds of assessment practices appropriate for a core curriculum. As changes are made in teaching practices within a core program, the methods employed to measure student achievement by districts, states, provinces, and student-certifying organizations such as the College Entrance Examination Board will be undergoing simultaneous changes. These kinds of issues should be raised early with planning and implementation groups and incorporated in the design of staff development on assessment.

Opportunity to Learn

At the heart of the concept of a core curriculum is the conviction that to differentiate learning goals prematurely by ability or by other means is to restrict students' opportunity to learn and to limit their futures. By the same token, restricting instruction within the core curriculum by beliefs about for whom technology is appropriate, by lack of funding for developmental materials, or by inadequate support structures for helping at-risk students also limits students' opportunity to learn.

A major aspect of implementing a core program is the education of adult constituencies about changes in the way that mathematics is used, its importance to students' futures, and the means required to prepare students for life in the twenty-first century. **It will absolutely take more resources.** As individuals and as a profession, we must press the argument that the character of mathematics learning no longer fits the pedantic memories of most adults. Mathematics is executed and applied in the real world in modernized ways that parallel the need to upgrade vocational education by providing appropriate equipment and technology to reflect real-world uses.

Learning activities and the means to assess them are assuming many characteristics of laboratory disciplines—multiple materials, different learning environments, expanded use of technology, and multiple forms of assessment. Improved support structures for assisting students to achieve are necessary for extending their opportunity to learn. Student needs must be addressed through the organizational and educational structure of the school as well as through classroom actions.

Just as a business invests heavily in training and upgrading the skills of its employees, education must invest similarly in the staff development needed to support the reform in mathematics programs. These are tough issues because they are costly issues. But policy shapers and the supporting public must be presented with these arguments and the ultimate consequences of not rising to the challenge of giving every student the opportunity to learn.

Leadership for Change

When change may affect many people, a major deterrent to starting is "chicken and egg thinking": "We can't do this before _____ happens." A more productive approach is, "We CAN do this *until* _____ happens!" Like any major change, the course of action for implementing a core curriculum will unfold *while* the vision is pursued. Careful thinking and advance planning are important. But a comprehensive plan of action does not have to be in place for a beginning to occur. Action begets more action. Inertia yields to enthusiasm and success. Problems that seem insurmountable at the outset often are solved simply through familiarity with the tasks.

Leadership is the essential key. Someone must take the helm and steer the effort constantly toward the goal. That person must understand not only the curricular and teaching implications of changes but also the *process* of change and how humans react to change.

Reaction to Change

The reality of change is that it is an adjustment—it requires a demarcation from a routine that we have settled into as somehow "right" or "normal." Accepting the change to a core curriculum will require many of us to alter our personal "construct" of how to teach students. This construct reflects our personal teaching philosophy, experiences as a student and as a professional, knowledge of subject matter, skills in human interaction, ability to work within constraints, and myriad other factors. Students and teaching *will be* different. How we judge these changes may bring us to grips with our acceptance of changed goals. One way to confront our beliefs is to list our perceptions of students and then to examine our instruction and programs in light of those beliefs.

Statement of Beliefs

We believe this of ALL students. They—

♦ may only *currently* be noncollege intending;

♦ will eventually hold some of the most influential and successful positions within their communities;

♦ are entitled to our best teaching efforts and interest regardless of their future career choices or interests;

♦ can best attain mathematical literacy and power through problem-centered, concept-oriented instruction;

♦ are engaged in a significant, opportunity-opening, respectable program of mathematical studies;

♦ differ in learning styles and orientation to mathematics rather than in understanding or motivation;

♦ possess similar human capacities and needs associated with learning—the need for success, curiosity, the willingness to learn, and response to motivation and praise;

♦ learn best by doing, are curious about real-world phenomena, tend to be now-oriented, have a low tolerance for ambiguity, and have learning strengths that have been largely untapped.

Although we believe the rewards more than justify the efforts, implementing a core curriculum cannot be expected to be an uninterrupted success story. Individuals experience stress when reacting to changes in their work, whether those changes are desired or not. New approaches, new programs, new "almost anything" seldom initially work as well as did the old way of doing things. New ways must be learned. Old habits inconsistent with the new must be discarded. Beliefs must accommodate changes in behavior. New means of coping must be found. How individuals react to the new roles expected of them will depend greatly on their ability to tolerate change.

Effective leaders will devise means to help individuals adjust to changes and understand the characteristics of change. They will understand the importance of continued feedback, looking for times and ways to reinforce individual efforts and encourage flagging resolve.

Some Common Aspects of Change

♦ Change interrupts an established pattern of doing things that seems and feels "right" because everyone shares a common set of expectations.

♦ More energy is required to do something differently. We must think through new actions to replace patterned ones that required little conscious thought.

♦ Stress occurs even in desired change because roles and expectations are uncertain until a new pattern is established and accepted. Efficiency and effectiveness often suffer until this occurs.

♦ Change is commonly resisted subconsciously as we try to make the change "fit into" the established mode of doing things, giving the illusion of making a change while maintaining the status quo.

♦ Changes in beliefs and expectations usually lag behind changes in behaviors. We behave differently, but we still expect past outcomes *as well as* those that should result from the change.

♦ Reversion to prior routines is an ongoing psychological need and a frequent action. Reversion may occur *even in light of superior performance,* to relieve the tension of change.

♦ Maladaptive change behaviors range from anxiety, confusion, apathy, and irritability to recounting the virtues of yesteryear.

Increased Sense of Community

A major benefit of changing to a core curriculum will be that staffs work together more professionally as units. Present programs sometimes permit teachers to become "specialized"—the honors geometry teacher, for example. Fracturing and splintering the curriculum into multiple sequences can have a similar fracturing and splintering effect on staff interest, focus, and responsibilities toward the whole program. In those instances, departments may act more like confederacies of loosely connected individualists than cooperative communities united by a common focus and concern for PROGRAM achievement.

Instituting a core program will bring department members into closer working relationships: more sharing of knowledge, more working together, and more vesting in expertise and dissemination. Individuals will need to give up some autonomy in making teaching decisions and engage in more cooperative planning, program assessment, and analyses of data. Those who have the knowledge and expertise needed to bring new topics to more diverse populations of students will need to share it. Where that expertise is not present, it will have to be sought in a systematic fashion. Inservice programs should become more focused and tied to goals and follow through. There should be a need; there should be an action; there should be a change. Staff members who go outside the district for professional growth should have a targeted department mission, as well as a personal agenda, to get the most out of such experiences.

Multiple refinements of core curricula are likely to continue to evolve into the future. The same forces rendering past skills obsolete will continue to influence expectations and our manner of learning mathematics. Therefore, it is important that staff members begin to understand change as an ongoing process. Implementing a core curriculum

The half-life of the education of an engineer has been estimated at ten years. In one decade, half of an engineer's training will become obsolete.

should not be perceived as a one-time event. Rather, it should be seen as the means for instituting a structure to oversee continuous curriculum change and to support individuals in adapting to those consequences.

CONCLUSION

The core curriculum—represented by the fourteen grades 9–12 standards and by their tenet of common outcomes for all students—is a curriculum concept with enormous power to revolutionize instruction and learning. It carries the potential for empowering all learners—but particularly those who traditionally have lacked access to challenging levels of ideas—to explore mathematics previously unattainable. Old teaching ideas must be modified and new approaches must be found to realize the potentials of the core. The roles of learner and teacher will change in exciting and revitalizing ways. Exactly how the changes will occur and the extent of those changes are not clear at the start. But the vista is so promising and the potential so rich that the journey must be joined.

Many issues will have been confronted and resolved by the time a core curriculum becomes institutionalized—money will have been found, resources expanded, and new physical and operational structures put into place to meet student needs. Materials will be developed to satisfy demands and teaching practices will be continually refined as we discover more effective ways to apply the core. The nonmathematics public will become increasingly convinced of the viability of the concept and increase their support.

For teachers, the changes in professional growth hold perhaps the most gratifying promise. Communicating among peers, sharing experiences, questioning beliefs, rising to the challenge of inspiring disenfranchised learners, and coming to new insights about the nature of mathematics and teaching will become new and well-traveled roads of our professional careers. The future of society and our hopes for meeting the challenges of tomorrow are more firmly in our hands than ever before.

Appendix I

MODEL SYLLABUS FOR A CROSSOVER CURRICULUM

The following syllabus for the Crossover Model corresponds to sample lessons in chapter 3. It is intended to be a transition to a core curriculum, paralleling existing coursework for the college-intending student. Except as noted in Year 3, all topics are considered core material for both the A and the B sequences of this model. The rearrangement of topics presents mathematics more as a way of modeling and describing real-world phenomena than as the more traditional deductive system. The major emphasis is on students constructing ideas through questioning, exploration, and discovery, with frequent use of models and applications. Patterning, graphing, inductive reasoning, and technology are major tools used to build ideas. De-emphasized are by-hand computation and algebraic manipulation skills.

General Suggestions for a Transition to a Crossover Core Curriculum

♦ Use current texts but reorder topics. Study the "Increased/Decreased Attention" suggestions in the *Curriculum and Evaluation Standards for School Mathematics* (NCTM 1989) and give important material more prominence. Often the most important ideas of a course are treated hurriedly at the end of the year. Put those topics up front to give more time for both development and review.

   • Begin with the more concrete topics (e.g., identification of shapes, properties, classifications, measurement work, real-world data).

   • Use inductive reasoning to develop ideas where deduction requires too many underpinnings. Consider postulating important chunks of content (e.g., using convincing informal means to demonstrate ideas to students), then use deductive reasoning from that base of understanding.

   • Insert transitional topics and supplement with new materials as necessary (e.g., matrices, coordinates, transformations, vectors). See the other volumes in the Addenda Series, Grades 9–12, for helpful suggestions.

   • Conclude coursework with modest deductive systems, giving care to the development of "if...then..." reasoning.

♦ Use questioning rather than telling whenever possible. All the sample lessons for this core model highlight a questioning approach. At a minimum, provide meaningful explanations of new material to students and ask them to write their understanding or explain to a partner. At best, encourage students to investigate questions of interest to themselves.

♦ Give applications more prominence to bring more relevance to ideas and to motivate reluctant learners.

   • Begin lessons with problem situations drawn from the students' world. Some situations may be drawn from later exercises for that lesson or from applications from supplemental text materials.

   • Begin sequences of related lessons and units with challenging problems that appear near the end of those instructional materials. Let students understand the problem, identify what is given and what is to be found, try simpler or related problems, and generally use heuristics to begin to grapple with ideas that they will study more explicitly during the unit. Problems can then be revisited during the unit(s) when more tools become available.

   • Have students write problems related to topics they have been studying. Take good application problems and omit the question, letting students complete the "setup." Create a problem bank for the class and select those problems most closely fitting the goals of the unit and integrate them into lessons, homework, or examinations.

♦ Include investigations in homework. Investigations may (a) extend earlier projects begun in class; (b) preview ideas needed for coming lessons; or

(c) relate to "umbrella questions" woven throughout the unit. Open-ended investigations allow students to pursue ideas in a manner and to a level that is comfortable. Students with a wide range of abilities can share the experience of understanding and grappling with a common problem and its associated mathematical ideas without the frustration of trying to achieve a preset level of performance.

◆ Maintain newly developed ideas throughout a course. Using traditional texts, include some problems from preceding lessons or the chapter review. Include some exercises that revisit important topics from previous chapters.

◆ Lighten the calculation and symbol manipulation load by integrating calculator and computer software use into all work.

◆ Evaluate some activities by holistic means (e.g., projects, problems of the week or the unit). Establish a benchmark for any student who puts in a solid effort toward an assigned task, then look for satisfactory, good, excellent, and outstanding work relative to this benchmark. For ideas, see *Assessment Alternatives in Mathematics* in the Annotated Bibliography of Core Resources.

SYLLABUS FOR THE CROSSOVER MODEL

Year 1—"Algebra"

Unit 1   *Patterns and Geometric Figures*

- ◆ Point, line, ray, plane, space, parallel, perpendicular, bisection, symmetry, polygon, circle
- ◆ Relations in space, informal loci
- ◆ Segment and angle measurement and congruence
- ◆ Construction and locus discoveries
- ◆ Angle relationships
- ◆ Polygons
- ◆ Predicting from patterns
- ◆ Informal notions of variable, evaluating formulas

Unit 2   *Exploring Data*

- ◆ Sorting and sequencing data
- ◆ Mean, median, variability
- ◆ Presentations of data: tables, matrices, frequency distributions, stem-and-leaf plots, line graphs, circle graphs, box plots, percentiles
- ◆ Dispersion, outliers, measures of dispersion

Unit 3   *Graphs*

- ◆ Coordinate system, ordered pairs, paired data, scatter diagrams
- ◆ Intuitive line fitting and interpretation of linear graphs
- ◆ Predicting rules for well-behaved linear data
- ◆ Graphing nonlinear paired data
- ◆ Using rules to produce linear graphs
- ◆ Graphing related families of lines, intuitive notions of slope and intercept
- ◆ Nonlinear graphs and paired data

Unit 4 *Algebraic Expressions*

- Writing and evaluating variable expressions
- Order of operations
- Generalizing number and geometric patterns and properties of numbers
- Equivalent and nonequivalent expressions
- Simplifying expressions

Unit 5 *Real Numbers*

- Models for integers and integer operations
- Absolute value, inequality, number line
- Properties of inequalities
- Solving simple linear equations

Unit 6 *Equations and Inequalities*

- Writing equations and inequalities for given conditions
- Solving linear equations and inequalities by equivalent operations
- Evaluating formulas
- Using tables to generate equations
- Solving problems that can be modeled by linear equations and inequalities
- Using graphs to solve linear equations and inequalities

Unit 7 *Matrices*

- Writing and interpreting information matrices
- Directed graphs, drawings, and networks
- Interpreting sums and products of matrices

Unit 8 *Properties of Geometric Figures*

- Exploration and organization of properties of quadrilaterals
- Properties of polygons and angles: triangle sum, polygon sum, exterior angles of polygons, angles of regular polygons
- Generation and solution of equations related to angle measures
- Area of parallelograms, triangles, and trapezoids
- Perimeter and area of polygons
- Circumference and area of circles

Unit 9 *Products and Polynomials*

- Products, quotients, and powers of monomials
- Multiplying and dividing a polynomial by a monomial, interpreting with area models
- Solving formulas
- Special products $(a + b)^2$ and $(a - b)(a + b)$ and area models
- Compound interest, population growth, depreciation, other exponential applications

Unit 10   *Rational and Irrational Numbers*

- Square roots, approximations
- Pythagorean theorem and applications
- Distance and midpoint formulas
- Applications involving squares and square roots
- Properties of real numbers: opposites, inverses, order, closure, identity, density
- Hierarchy of real-number subsets

Unit 11   *Variation and Graphs*

- Ratio and proportion applications
- Variations: direct, inverse, square, joint
- Interpretations of direct and inverse variation graphs in terms of rates, constants of variation, effect of parameter changes on graphs
- Writing formulas to express variation relationships given in written, tabular, and graphical form (implicit slope definition)

Unit 12   *Functions and Relations*

- Solutions of equations in two variables
- Graphing relations
- Input-output model of a function
- Slope-intercept form of linear functions, parallel and perpendicular lines
- Families of lines: m as a stretcher, b as a translation
- Writing equations for lines satisfying given conditions
- Graphing linear inequalities

Year 2—"Geometry"

Unit 1   *Shapes in Space*

- Introduction to tools of geometry: models, protractors, compasses, algebra, heuristics
- Describing 1-, 2-, 3-D figures
- Drawing and sketching 2-, 3-D figures
- Angle relationships
- Symmetry

Unit 2   *Patterns in Shapes*

- Triangle sum property leading to classification of triangles
- Polygon sum property, exterior angle property, angles in regular polygons
- Introduction to inductive versus deductive reasoning and counterexamples
- Properties of quadrilaterals leading to hierarchy of quadrilaterals
- Inscribed angle property

Unit 3   *Measurement in the Plane*

- Distinctions among 1-, 2-, and 3-D measures
- Perimeter and circumference in the plane

- Areas of squares, rectangles, parallelograms, triangles, trapezoids
- Pythagorean theorem
- Special right triangles
- Geometric probability using 2-D models

Unit 4 *Shapes on the Coordinate Plane*

- Vectors
- Slope
- Parallels and perpendiculars on the plane
- Midpoint and distance formulas
- Coordinates for triangles and quadrilaterals
- Deductions based on coordinates

Unit 5 *Transformations*

- Reflections over axes and $y = x$
- Translations, vectors
- Rotations by multiples of $90°$
- Size changes $(x, y) \rightarrow (ax, ay)$
- Scale changes $(x, y) \rightarrow (ax, by)$
- Invariants
- Definitions and uses of congruence and similarity
- Tessellations

Unit 6 *Measurement in Space*

- Surface area and volume of prism and cylinder
- Surface area and volume of pyramid and cone
- Surface area and volume of sphere
- Geometric probability using 3-D models
- Similar figures in space and ratios of similarity for 1-, 2-, and 3-D measures

Unit 7 *Trigonometry*

- Right triangle properties (constant ratios)
- Applications of trigonometric ratios
- Unit semicircle
- Law of sines
- Patterns for unique triangles: *ASA, AAS*
- Law of cosines
- Patterns for unique triangles: *SAS, SSS*
- Ambiguous patterns: *AAA, SSA*

Unit 8 *Logic*

- Statements and negations
- Conjunctions and disjunctions
- Implications and their translations
- Converses
- Properties as implications
- Double implications (biconditionals)
- Definitions as double implications
- Flow proofs

Unit 9 *Reasoning with Angles and Parallel Lines*

- A minimal deductive system for typical proofs about supplements, complements, angles formed by parallel lines

Unit 10 *Reasoning with Triangles*

- A minimal deductive system for proofs about congruent and similar triangles (e.g., earlier work with patterns for unique triangles—SSS, and so on—are postulated)

Year 3—"Advanced Algebra"

(Δ signifies a B-sequence topic only)

Unit 1 *Patterns and Properties of Numbers*

- Extending and generalizing patterns of numbers, arrays, geometric figures
- Sequences and series
- Recursive formulas
- Finite differences
- "Explaining" number tricks
- Field properties

Unit 2 *Linear Equations and Inequalities*

- Solving linear equations and inequalities
- Slope-intercept form of linear equations; parallel/ perpendicular lines
- Absolute value
- Literal equations and formulas
- Logical connectives: and, or
- Solution of systems of linear equations by graphing, tables, substitution
- Solution of systems of linear inequalities by graphing
- Linear programming
- Δ Absolute value inequalities
- Δ Solution of systems of linear equations by addition

Unit 3 *Quadratic Equations and Relations*

- Quadratic functions
- Solving quadratic equations
- Solving quadratic equations by graphing, quadratic formula
- Conics with center at the origin: locus definitions, optical properties, applications
- Δ Completing the square
- Δ Discriminant
- Δ Solution of quadratic equations by factoring

Unit 4 *Functions and Graphs*

- Functions and relations, input-output model, function notation
- Step functions
- Graphical introduction of exponential, logarithmic, sine, cosine functions

- ◆ Characteristics of graphs: symmetry, max-min, increasing-decreasing, excluded regions, periodicity
- ◆ Translations, effects of parameter changes on linear and quadratic relations and functions
- △ Graphical solution of nonlinear systems

Unit 5   *Polynomials and Polynomial Functions*

- ◆ Exponent laws (integer, $1/n$)
- ◆ Sum, product of monomials and polynomials
- ◆ Pascal's triangle
- ◆ Graphing polynomial functions
- ◆ Roots by graphical estimation
- △ Remainder, factor theorems
- △ Fundamental theorem of algebra
- △ Inverse functions

Unit 6   *Matrices*

- ◆ Information matrices
- ◆ Directed graphs, drawings, and networks
- ◆ Sums and products of matrices
- ◆ Transformations of geometric figures by matrices
- △ Inverse of a square matrix, solution of linear systems

Unit 7   *Real and Complex Number Systems*

- ◆ Rational and irrational numbers
- ◆ Rational exponents
- ◆ Real numbers and field properties
- ◆ Imaginary numbers
- ◆ Operations with complex numbers
- ◆ Hierarchy of complex number subsets
- △ Finite systems
- △ Radical equations
- △ Algebraic proof

Unit 8   *Exponential and Logarithmic Functions*

- ◆ Graphing exponential and logarithmic functions
- ◆ Exponential growth and decay, applications and modeling
- ◆ Logarithmic applications
- ◆ Solution of simple exponential and logarithmic equations
- △ Properties of exponents and logarithms
- △ Simplifying exponential and logarithmic expressions

Unit 9   *Probability and Statistics*

- ◆ Counting situations, tree diagrams
- ◆ Simple probability experiments and applications
- ◆ Theoretical probability and simulations
- ◆ Sampling techniques, random numbers
- ◆ Monte Carlo techniques
- ◆ Mean and standard deviation
- ◆ Normal distribution

△ Conditional probability

△ Permutations and combinations

△ Binomial theorem

△ Linear regression and correlation

Unit 10 *Sequences, Series, and Limits*

♦ Arithmetic sequences

♦ Geometric sequences

♦ Arithmetic and geometric series

♦ Infinite geometric series

△ Sigma notation

△ Secant and tangent lines, notion of limiting behavior

△ Area under a curve

Unit 11 *Circular Functions*

♦ Sine and cosine graphs and periodic behavior

♦ Degree and radian measures

♦ Applications of periodic motion

♦ Solving simple equations involving real-world phenomena

△ Tangent function and graph

△ Graphical properties of $y = a \sin bx$ and $y = a \cos bx$

Unit 12 *Rational Expressions and Functions*

♦ Simplifying, multiplying, adding rational expressions

♦ Fractional equations and applications

♦ Inverse variation

♦ Graphing simple rational functions

♦ Asymptotes

△ Rational functions and horizontal and vertical asymptotes

Year 4—"Advanced Mathematics"

(All topics are for college-intending students)

Unit 1 *Difference Equations*

♦ Finite graphs and applications

♦ Algorithms and recurrence relations

♦ Programming operations (initialize, increment, test, systematic search)

Unit 2 *Operating with Functions*

♦ Modeling with—
  • compositions of functions created from simpler functions
  • piecewise functions
  • rational, radical, transcendental functions

♦ Limits

♦ End-behavior

Unit 3 *Functions and Equations*

- Theory of equations (factor/remainder theorems with symbol manipulator/graphing software)
- Equations with radicals (one-way and reversible steps)

Unit 4 *Circular Functions*

- Connection of trigonometric and circular functions
- Modeling, graphing of circular functions
- Inverses
- Trigonometric equations
- Proving identities

Unit 5 *Vectors*

- Applications to real situations
- Coordinate form $[a, b]$ and polar form (r, θ)
- Algebraic construction and application

Unit 6 *Applications of Matrices*

- Matrix transformations
- Solution of linear systems
- Linear programming

Unit 7 *Complex Numbers and Polar Coordinates*

- Matrices for complex numbers
- Matrix multiplication and complex number operations
- DeMoivre's theorem

Unit 8 *Probability and Statistical Inference*

- Transformations of linear and exponential data
- Hypothesis testing
- Random variables to generate and interpret probability distributions

Unit 9 *Advanced Proof Ideas*

- Indirect proof
- Mathematical induction
- Comparisons among various axiomatic systems (alternative geometries)
- Nature and purpose of axiomatic systems

Unit 10 *Rates and Areas*

- Average and instantaneous rates of change
- Secants and tangents
- Derivative from the definition
- Areas by grid, Monte Carlo, rectangle, and trapezoid methods
- Integrals from the definition

Appendix II

MODEL SYLLABUS FOR A DIFFERENTIATED CURRICULUM

The following syllabus for the Differentiated Model corresponds to sample lessons in chapter 4. It represents a single-sequence core curriculum in which differentiation of instruction is expected to occur within units as determined by students' needs. The first three years of the four-year sequence constitute the core, which all students are expected to experience. The unified nature of this curriculum provides a robust framework for achieving all fourteen core standards for grades 9–12 as presented in the *Curriculum and Evaluation Standards for School Mathematics* (NCTM, 1989), but most particularly for achieving those standards involving problem solving, communication, reasoning, and connections. This syllabus lists unit topics and the major standard sections addressed, including comments, to enable the reader to better ascertain how unit topics are expected to be treated. Code: *m.n* refers to standard *m,* bullet *n* in the grades 9–12 Standards.

SYLLABUS FOR THE DIFFERENTIATED MODEL

Year 1—Patterns and Properties

Unit 1   *Exploring Geometric Figures*

   Topics:

- Basic vocabulary and, where appropriate, symbolism for *point, line, ray, plane, parallel, perpendicular, bisection, symmetry, polygon, circle, polyhedron,* and the terms associated with each
- Relations in space, informal loci, 3-dimensional sketches
- Angle relationships
- Symmetries and lines of reflection
- Locus discoveries using constructions
- Inductive reasoning, conjecture, counterexample, convincing argument

Major standards addressed: 2.2, 3.1, 3.2, 3.3, 3.5, 7.1, 7.2, 7.4

   Comments:

- Very exploratory and informal
- Introduces vocabulary and symbols as need arises
- Assumes measurement skills from grades 5–8
- Uses geometric software for exploring and conjecturing about properties of lines, angles, polygons
- Includes historical perspectives

Unit 2   *Exploring Data*

   Topics:

- Presentations of data: tables, matrices, frequency distributions, stem-and-leaf plots, line graphs, circle graphs, box plots, percentiles
- Measures of central tendency
- Dispersion, outliers, measures of dispersion

Major standards addressed: 4.4, 10.1, 10.3, 12.1

Comments:

- Uses statistical software and spreadsheets to display data and measures of central tendency and dispersion
- Has data stemming from student experiments and sources related to sports, entertainment, and other academic courses

Unit 3   *Graphs*

Topics:

- Scales, particularly in computer/graphics calculator windows
- Distances on the number line and absolute value
- Operations with integers with and without a number line
- Representations of solutions to open sentences and inequalities on a number line
- Representations of ordered pairs on the coordinate plane
- Exploration of numeric, geometric, and situational sequences that build to formulas including those representing patterns that are linear, quadratic, inverse, exponential
- Explorations of the "shapes" of formulas with function graphing software (informal introduction to function families)

Major standards addressed: 5.1, 5.2, 6.2

Comments:

- Provides visual representation of formulas students have generated
- Introduces to graphing utilities after students have had experience graphing basic functions ($y = ax$, $y = x + b$, $y = x^2$, $y = 1/x$, $y = |x|$) by hand

Unit 4   *Expressions, Sentences, and Situations*

Topics:

- Continued use of numeric, geometric, and situational sequences that build to formulas, particularly those representing patterns that are linear
- Exploration of expressions that may or may not be equivalent, emphasizing the distributive property
- Creation and evaluation of formulas
- Informal solution of sentences generated from word situations and applications
- Introduction to the idea of explicit and recursive formulas

Major standards addressed: 5.1, 5.2, 6.2, 6.3, 12.1

Comments:

- Introduces variable as a generalizer for numeric patterns
- Uses tables and systematic guess and check
- Includes work in reading and modifying simple computer programs to generate the patterns discovered
- Nonequivalent expressions suggest need for agreement on order of operations.
- Equivalent expressions become basis for properties
- Symbolism, such as subscripts, is used as need arises
- Foreshadows work with special types of functions

Unit 5 *Models for Operations*

Topics:

- Characteristics of situations requiring addition (putting together, slides), subtraction (take away, missing addend, comparison), multiplication (arrays, areas, rate factor, magnification/reduction), division (ratios, rates, missing factor, partitioning, successive subtraction)

- Modeling situations using the operations

- Properties of the four basic operations

- Undoing the operations and solving and checking solutions to open sentences of the forms $ax = b$, $x + b = c$, $ax + b = c$

Major standards addressed: 5.1, 5.2, 14.2

Comments:

- Situations for each operation are done in tandem with the properties for the operation.

- Realistic situations with fractions, decimals, and percents occur throughout.

- Includes ratios and proportions

- Situations lead to sentences that suggest the need for procedures.

Unit 6 *Linear Situations, Sentences, and Graphs*

Topics:

- Solving problems by generating equations of the form $ax + b = cx + d$ and forms with parentheses

- Properties of inequalities

- Solving problems by generating linear inequalities

- Graphs of the forms $y = ax$, $y = x + b$ with recognition of the roles of the parameters

- Graphs of equations and inequalities of general linear form $y = mx + b$ ($y \leq mx + b$)

- Finding equations for lines meeting certain conditions

Major standards addressed: 5.1, 5.2, 5.3, 6.2, 6.4, 6.5

Comments:

- Situations generate sentences that suggest the need for procedures.

- Graphing utilities produce various functions of the form $y = ax$ or $y = x + b$ so that students can infer roles of parameters.

- An informal transformation approach is used for properties of inequalities.

- Adding or subtracting the same value slides the graph and preserves the original inequality.

- Slopes and intercepts for lines are interpreted in terms of the real situation context from which they arise.

Unit 7 *Products and Powers*

Topics:

- Compound interest, population growth, depreciation, and other situations giving rise to exponential growth

- Introduction to symbol manipulator

- Products of powers and quotients

- Powers of products, quotients, and powers
- Area models for $(a + b)^2$ and $(a + b)(a - b)$
- Other polynomial products with area or volume interpretations

Major standards addressed: 5.1, 6.2, 6.4, 14.1, 14.2

Comments:
- Properties of powers are inferred from hand calculations and from observation of results on symbol manipulators.
- Some work with powers is done mentally.

Unit 8 *Special Powers*

Topics:
- Negative exponents
- Square roots
- Powers of $1/n$
- Properties of roots
- Pythagorean theorem proved and applied in multiple ways
- Special right triangles
- Hierarchy of number systems leading to real numbers

Major standards addressed: 5.1, 5.3, 6.1, 6.2, 6.4, 7.2, 7.4, 14.1, 14.2

Comments:
- Pythagorean theorem is proved in multiple ways.
- Properties of roots arise through numerical investigations as well as by inferences from a symbol manipulator.

Unit 9 *Properties of Geometric Figures*

Topics:
- Exploration and organization of properties of quadrilaterals
- Properties of polygons and angles: triangle sum, polygon sum, exterior angles of polygons, angles in regular polygons
- Generation and solution of equations related to angle measures
- Congruence using rigid motions and similarity using magnification of images

Major standards addressed: 7.2, 7.3, 7.4

Comments:
- Applies linear equations regularly
- Contrasts inductive reasoning (looking at many specific cases to form a generalization) with deductive reasoning (arguing from a general case)
- Identifies many situations where congruence and similarity arise
- Informally approaches congruence and similarity by means of folds, slides, rotations, and synthetic size transformations

Unit 10 *Measures in Geometry*

Topics:
- Circumference and arc length
- Areas of irregular figures
- Areas of triangles and trapezoids

- Areas of kites
- Areas of circles and sectors
- Volumes of right prisms and cylinders
- Volumes of pyramids and cones
- Surface area and volume of spheres

Major standards addressed: 7.2, 7.4

Comments:

- Assumes perimeters of polygons, areas of squares and rectangles, and volumes of cubes and rectangular parallelepipeds
- Irregular areas are done by "breaking down" the figure into manageable parts and overlaying grids.
- Emphasizes the dimension required in a situation (e.g., amount of fabric needed to cover a couch [2-D] versus the amount of stuffing needed to fill the couch [3-D]
- Is organized so that students see simple patterns (e.g., arc length : circumference :: sector area : circle area; pyramid : prism :: cone : cylinder)
- Deduces with models for areas of triangles, trapezoids, and kites
- Uses sand or vermiculite to dramatize volume relationships
- Continues to solve equations and inequalities using geometric measurement situations

Unit 11   *Introduction to Probability and Simulation*

Topics:

- Definition of theoretical and empirical probabilities
- Randomness and uniform distribution
- Geometric probability using 1-, 2-, and 3-D models
- Design and execution of simple experiments

Major standards addressed: 10.1, 10.4, 10.5, 11.1, 11.2, 11.3, 11.4

Comments:

- Begins with simple experiments as a vehicle for introducing terminology
- Frequently compares theoretical and empirical results
- Examines random number generators and observes tendency toward uniformity of distribution as trials increase
- Builds on geometric measurement from previous unit

Unit 12   *Introduction to Functions*

Topics:

- Definition of function
- Systems solved by using graphing utility
- Situations modeled by quadratic growth
- Identification of maxima/minima by using graphing utility
- Solving equations by using graphing utility: inverse, quadratic, square root, exponential

Major standards addressed: 5.1, 5.2, 6.1, 6.2, 6.3, 6.4

Comments:

- *Function* notation is suggested by the need to discuss two or more functions at the same time.

- Graphing utility is used to find intersections to solve systems.
- The function $f(x) = k$ is solved by finding intersection of the graph of $y = f(x)$ and the horizontal line $y = k$.
- Serves as precursor to extended work with functions in later courses
- Provides numerous situations for each type of function

Year 2—Visualizing Relationships

Unit 1   Variation and Modeling

Topics:
- Introduction to modeling
- Situations, tables, graphs, equations for variations: direct, inverse, square, inverse square
- Finite differences for quadratic models
- Deducing from a model and inferring a model from data

Major standards addressed: 1.4, 5.1, 5.2, 6.1, 6.2, 6.3, 6.4, 10.1, 10.2, 13.1

Comments:
- Connects situations of various types to variation models
- Continues to interpret parameters of equations in terms of the original situation
- Illustrates the idea of holding all but one variable constant in order to recognize the type of variation for one pair of variables
- Includes forming models from data as well as predicting from determined models
- Uses curve-fitting software

Unit 2   Coordinate Geometry

Topics:
- Distance and midpoint formulas
- Slopes of parallel and perpendicular lines
- Proofs with respect to the coordinate plane

Major standards addressed: 7.4, 8.1, 8.2, 8.3, 8.4

Comments:
- Includes vectors as ordered pairs
- Assumes slopes from previous work
- Numeric cases lead to variable.
- Continues to build idea of inductive versus deductive reasoning

Unit 3   Transformations of Geometric Figures

Topics:
- Reflections and symmetry
- Translations and vectors
- Rotations of multiples of 90°
- Size transformations and two-way stretches
- Composites of transformations

- Invariant properties of geometric figures under various sets of transformations
- General definitions of congruence and similarity
- Applications of congruence and similarity

Major standards addressed: 8.1, 8.2, 8.3, 8.4

Comments:
- Emphasizes transformations and coordinates
- Applies distance, midpoint, and slope formulas to investigate invariant properties
- Defines congruence in terms of images resulting from isometries or their composites
- Defines similarity in terms of images resulting from isometries, size transformations, or their composites
- Two-way stretch provides contrast to size transformation; the former does not preserve ratios.
- Applications include ratios of perimeters, surface areas, and volumes under size transformations.

Unit 4   *Introduction to Trigonometry*

Topics:
- Ratios for right triangles
- Applications of right triangle trigonometry
- Unit circle (quadrants I and II)
- Laws of sines and cosines
- Drawing of oblique triangles meeting given conditions and verification of measures using laws of sines and cosines
- Recognition of triples (e.g., *SAS, ASA*) that ensure unique results

Major standards addressed: 7.4, 9.1

Comments:
- Unit circle is done with calculator and graph grid to make concrete the meanings of sine and cosine.
- Students draw oblique triangles and confirm by trigonometry exact solutions, impossibilities, or multiple solutions.
- This approach to congruence avoids the more lengthy and less concrete traditional treatment.

Unit 5   *Functions*

Topics:
- Concept of unique output for one input
- Absolute value, step, and constant functions
- Graphs and situations, working back and forth
- Lines and parabolas revisited as functions
- Quadratic formula and graphic interpretation
- Informal exploration of transformations of functions

Major standards addressed: 6.1, 6.2, 6.3, 6.4, 10.1, 10.2

Comments:
- Generalizes earlier work

- Includes situations using various functions
- Treatment of transformations is visual and verbal, not symbolic; focus is on seeing and describing what stays the same, what changes and how it changes.

Unit 6 *Lines, Parabolas, and Exponential Curves*

Topics:
- Linear and geometric sequences (explicit and recursive definitions) and their graphs
- Quadratic growth as distinct from linear and exponential growth
- Complex numbers and solutions to all quadratics
- General forms for linear, quadratic, and exponential functions
- Exponential modeling

Major standards addressed: 6.1, 6.2, 6.3, 6.4, 6.5, 12.1

Comments:
- Summarizes and generalizes earlier ideas
- Emphasizes quadratic and exponential functions and situations giving rise to them

Unit 7 *Transformations of Functions and Data*

Topics:
- Translations of data and effects on graphs (e.g., box plot, histogram), measures of central tendency, and dispersion
- Scale changes (one-way stretch) of data and effects on graphs, measures of central tendency, and dispersion
- Standard scores as composite of translation by the mean and scale change by the standard deviation, applications to comparisons
- Translations of functions and effects on graphs and equations
- Scale changes of functions and effects on graphs and equations

Major standards addressed: 6.5, 10.2, 10.6

Comments:
- Data transformations build directly from geometric work in earlier chapters.
- Standard scores can be applied to ACT and SAT scores and to other examples of student interest.
- Function transformations are consistently handled from $y = f(x)$ form so that any substitution (e.g., ay for y) results in the inverse operation in the equation $y/a = f(x)$.
- A major goal is for students to be able to readily recognize the reasonableness or unreasonableness of a graph displayed by a graphing utility.

Unit 8 *Systems*

Topics:
- Graphic solution of systems
- Linear programming graphically

Major standards addressed: 5.1, 5.2, 5.3, 6.1, 6.2, 6.3

Comments:

- Allows opportunities to continue to apply modeling concepts for all functions previously studied and transformation concepts from the previous chapter
- Uses tables and spreadsheets in addition to other methods

Unit 9   *Matrices*

Topics:

- Matrices for storage of information
- Interpretations of sums and products of matrices
- Networks, their graphs, and connectivity matrices

Major standards addressed: 12.1, 12.2

Comments:

- Emphasizes problem-solving strategies; very visual and intuitive
- Includes use of spreadsheets
- Supplies a motivation for multiplication of matrices from business, inventory, and sales situations
- Uses technology for operating with large matrices.

Unit 10   *Combinatorics and Binomial Distributions*

Topics:

- Permutations and combinations derived from counting situations
- Pascal's triangle and its properties
- Binomial theorem
- Simple applications to probability
- Simple binomial experiments with comparisons between theoretical and empirical results

Major standards addressed: 11.1, 11.2, 11.4, 12.4

Comments:

- Emphasis on problem-solving strategies leads students to their own formulas.

Year 3—Functions and Reasoning

Unit 1   *Fitting Curves to Data*

Topics:

- Linear regression
- Correlation
- Fitting polynomials to data
- Issues of errors in measurement
- Sampling considerations
- Fitting exponential curves to data

Major standards addressed: 1.4, 6.1, 10.1, 10.2, 10.3, 10.4, 10.5

Comments:

- Connects many topics from students' earlier work: sampling, measurement, table generation, recognizing patterns typical of functions, using transformations to find a function to fit a data set
- Uses technology for curve fitting

Unit 2   *Circular Functions and Models*

Topics:

- Situations giving rise to sinusoidal functions
- Graphs of circular functions
- Exploration of characteristics of circular functions versus conditions in situations generating them
- Properties of circular functions from their graphs

Major standards addressed: 9.2, 13.1

Comments:

- This topic is done in an exploratory way and is not completely formalized, as suggested in the *Curriculum and Evaluation Standards.*

Unit 3   *Exponential and Logarithmic Functions*

Topics:

- Informal introduction to domain and range
- Compositions of functions through genealogy functions (e.g., is the father of, the father of the mother)
- Reflections over $y = x$ and inverses of functions
- Exponential and logarithmic functions
- The number e
- Situations giving rise to logarithmic growth
- Modeling exponential or logarithmic data by using log x or log y

Major standards addressed: 6.1, 6.2, 6.3

Comments:

- Continues the modeling thread with exponential and logarithmic data
- Logarithmic work is done with tables, graphs, and technology.

Unit 4   *Logic*

Topics:

- Translations of implication
- Valid arguments: direct argument, chain rule, indirect argument
- Biconditionals and definitions
- Statements with quantifiers
- Negations
- Inferential logic used in sampling

Major standards addressed: 2.2, 3.4, 3.5

Comments:

- Looks back to inductive and deductive distinctions made in earlier courses

- Looks back to inferential nature of statistical reasoning
- Uses examples from advertising, legal affairs, club rules, and other relevant situations

Unit 5 *Reasoning in Geometry*

Topics:
- Deductions in limited domains (e.g., parallels and similarity)

Major standards addressed: 3.4, 3.5, 7.2, 7.4

Comments:
- May use geometry tutoring software to assist students in developing proofs
- Includes invalid deductions (converse and inverse errors) for which students can give counterexamples

Unit 6 *Reasoning in Algebra*

Topics:
- Algebraic identities
- Deductions from natural numbers, integers, rationals, reals
- Logic of algebraic procedures
- Finite systems and their isomorphisms

Major standards addressed: 3.1, 3.2, 3.3, 3.4, 3.5, 14.1, 14.2, 14.3

Comments:
- Revisits some familiar content by applying the logic of earlier chapters
- Includes attention to equation-solving procedures that are not reversible (one-way implications) and connects the logic to algebra (extraneous roots or loss of roots)

Unit 7 *Reasoning in Intuitive Calculus*

Topics:
- Sequences, series, functions, and limits
- Successive approximation techniques
- Average and instantaneous rates of change
- Slopes of secants approaching slopes of tangents
- Area under a curve

Major standards addressed: 13.2, 13.3

Comments:
- Builds notions of real numbers from the previous chapter by using limits
- Slope of a tangent line may be informally explored as the slope of a graph itself by repeatedly zooming in with a graphing utility until the curve becomes a line.
- Similarly, functions without defined derivatives at a point can be zoomed in on from both sides to illustrate two different slopes.

Unit 8 *Reasoning in Discrete Mathematics*

Topics:
- Situations represented by finite graphs and deductions from them

- Algorithm development and analysis
- Recurrence relations and deductions from them
- Aspects of programming (e.g., initializing, incrementing, testing, and systematic searching)

Major standards addressed: 12.1, 12.2, 12.3

Comments:
- Continues to use tables, graphs, and verbal explanations to reduce difficulties with formal symbolism

Unit 9   *Reasoning in Probability*

Topics:
- Sets, unions, and intersections
- Counting problems, "or" and "and"
- Theoretical probability
- Conditional probability
- Review of permutations and combinations
- Binomial experiments

Major standards addressed: 11.1, 11.4

Comments:
- Furnishes ample practice making complete lists or generating (and implementing) algorithms for making such lists so that students develop intuition about counts before properties of probabilities are discussed

Unit 10   *Reasoning in Statistics*

Topics:
- Sampling concerns
- Monte Carlo techniques
- Normal distributions

Major standard addressed: 11.5

Comments:
- Normal distribution as derived from the binomial builds from earlier work with limits.
- Situations are amply modeled by normal distributions.
- Data bases are pre-entered on disk so students can easily generate frequency distributions, observe shape of data, and overlay and interpret means and standard deviations.

Year 4—Mathematics for the College-intending Student

Unit 1   *Operating with and Describing Functions*

Topics:
- Functions with piecewise rules
- Operations with functions (e.g., sum, quotient) giving rise to rational functions
- Domain and range
- Intervals of increase and decrease
- Horizontal and vertical asymptotes
- Graphical idea of limit

- Limit notation
- End behavior
- Oblique asymptotes

Major standards addressed: 6.1, 6.2, 6.4, 6.6, 13.1, 13.4

Comments:
- Emphasis is on a visual, graphic approach.
- Limit notation is connected to graphic representation.
- Continues to build from real situations, generating new functions
- Connects asymptotes and division

Unit 2 *Functions and Equations*

Topics:
- Polynomial, rational, radical, and transcendental functions and equations
- Factor and remainder theorems
- Fundamental theorem of algebra
- Nonreversibility of steps (converse error) in solving certain equations

Major standards addressed: 5.5, 6.3, 13.4

Comments:
- Continues to use background of transformations of functions to identify invariants and to predict changes
- Connects, wherever possible, graphs with equations and these representations with real situations

Unit 3 *Circular Functions*

Topics:
- Trigonometric identities
- Inverses
- Solving equations with and without a graphing utility
- Modeling with trigonometric functions

Major standards addressed: 9.3, 9.4, 9.5, 9.6

Comments:
- Trigonometric sentences can be graphed to conjecture which are identities and which are open sentences.
- Presents formal modeling rather than exploring for comparable objectives as in previous courses

Unit 4 *Applications of Matrices*

Topics:
- Vectors and transformations
- Matrices and transformations
- Dependence of vectors
- Linear programming
- Solutions of systems through matrices
- Basis of a system
- Matrices for general rotations in the plane

Major standards addressed: 5.4, 8.5, 8.6, 12.5

Comments:
- Uses calculators and computer software to manipulate matrices
- Interprets vector addition, scalar product, and dot product in terms of translation, scale transformation, and link to cosine of angle between two vectors

Unit 5   *Complex Numbers and Polar Coordinates*

Topics:
- Relationship of series and trigonometric functions
- Complex numbers
- Matrices for complex numbers and operations
- Polar coordinates
- Polar graphs
- DeMoivre's theorem
- Isomorphisms of trigonometric/polar forms

Major standards addressed: 9.7, 14.4

Unit 6   *Recursion*

Topics:
- Iteration and recursion
- Sigma notation
- Sums of sequences
- Limits of geometric series
- Difference equations

Major standards addressed: 12.5, 13.2, 13.3

Unit 7   *Advanced Proof Ideas*

Topics:
- Indirect proof
- Vector proofs
- Non-Euclidean geometry
- Sigma proofs including statistics formulas
- Proof by mathematical induction
- Comparisons among proof systems (e.g., synthetic versus analytic proofs of geometry theorems)

Major standards addressed: 3.6, 7.5, 8.5, 14.6

Unit 8   *Rates and Areas*

Topics:
- Average and instantaneous rates of change
- Secants approaching tangents
- Derivation of general derivatives from the definition
- Areas through grids, Monte Carlo methods, rectangles, or trapezoids
- Derivation of integrals from the definition
- Visualizing volumes of revolution

Major standards addressed: 13.2, 13.3

Comments:

- Includes recognition of the relationships of instantaneous rate and area under a curve for a few functions that are investigated in detail
- Volumes are determined informally through spinning cardboard models and 3-D graphing software.

Unit 9   *Statistical Inference*

Topics:

- Central limit theorem
- Hypothesis testing
- Tests for binomial and normal distributions, Poisson distribution, student's t test, chi square distribution
- Some nonparametric techniques
- Design of experiments
- Types of errors

Major standards addressed: 10.6, 10.7, 10.8, 11.5, 11.6

Unit 10   *Algebra and Algorithms*

Topics:

- Group and field postulates
- Proofs for groups and fields
- Isomorphisms of some systems
- Algorithms
- Computer validation of algorithms

Major standards addressed: 12.6, 14.3, 14.5, 14.6

Comments:

- Includes a historical perspective connecting the linked roots of algebra and algorithms

REFERENCES

American Association for the Advancement of Science. *Science for All Americans: A Project 2061 Report on Literacy Goals in Science, Mathematics, and Technology.* Washington, D.C.: The Association, 1989.

Blackwell, David, and Leon Henkin. *Mathematics: A Project 2061 Panel Report.* Washington, D.C.: American Association for the Advancement of Science, 1989.

Burrill, Gail, John C. Burrill, Pam Coffield, Gretchen Davis, Jan de Lange, Diane Resnick, and Murray Siegel. *Data Analysis and Statistics across the Curriculum.* Curriculum and Evaluation Standards for School Mathematics Addenda Series, Grades 9–12. Reston, Va.: National Council of Teachers of Mathematics, forthcoming.

California State Department of Education. *A Question of Thinking: A First Look at Students' Performance on Open-ended Questions in Mathematics.* Sacramento, Calif.: The Department, 1989.

Coxeter, H. S. M. *Introduction to Geometry.* 2d ed. New York: John Wiley & Sons, 1969.

de Lange, Jan. *Matrices.* Pilot instructional unit, Whitnall High School, Greenfield, Wisc., 1990.

Demana, Franklin B., and Joan R. Leitzel. *Essential Algebra: A Calculator Approach.* Reading, Mass.: Addison-Wesley Publishing Co., 1989.

Froelich, Gary W., Kevin G. Bartkovich, and Paul A. Foerster. *Connecting Mathematics.* Curriculum and Evaluation Standards for School Mathematics Addenda Series, Grades 9–12. Reston, Va.: National Council of Teachers of Mathematics, 1991.

Gleick, James. *Chaos: Making a New Science.* New York: Penguin Press, 1988.

Green, Philip C., et al. *Algebra and Statistics.* Dayton, Ohio: EFA and Associates, 1988.

Haldane, J. B. S. "On Being the Right Size." In *The World of Mathematics,* Vol. 2. Edited by James E. Newman, pp. 952—57. New York: Simon and Schuster, 1956.

Hirsch, Christian R., ed. *Activities for Implementing Curricular Themes from the "Agenda for Action."* Reston, Va.: National Council of Teachers of Mathematics, 1986.

Hirsch, Christian R., and Harold L. Schoen. "A Core Curriculum for Grades 9–12." *Mathematics Teacher* 82(December 1989): 696–701.

Hord, Shirley M., et al. *Taking Charge of Change.* Alexandria, Va.: Association for Supervision and Curriculum Development, 1987.

Johnston, William B., and Arnold E. Packers. *Workforce 2000: Work and Workers for the Twenty-first Century.* Indianapolis, Ind.: Hudson Institute, 1987.

Kenelly, John W., ed. *The Use of Calculators in the Standardized Testing of Mathematics.* New York: College Entrance Examination Board, 1989.

Madsen-Nason, Anne, and Perry E. Lanier. *Pamela Kaye's General Math Class: From a Computational to a Conceptual Orientation.* Research Series No. 172. East Lansing, Mich.: Michigan State University, 1986.

Mathematical Sciences Education Board. *Everybody Counts: A Report to the Nation on the Future of Mathematics Education.* Washington, D.C.: National Academy Press, 1989.

———. *Reshaping SCHOOL Mathematics: A Philosophy and Framework for Curriculum.* Washington, D.C.: National Academy Press, 1990.

National Council of Teachers of Mathematics. *An Agenda for Action: Recommendations for School Mathematics of the 1980s.* Reston, Va.: The Council, 1980.

———. *Curriculum and Evaluation Standards for School Mathematics.* Reston, Va.: The Council, 1989.

_____. *Discrete Mathematics across the Curriculum, K–12.* 1991 Yearbook, edited by Margaret J. Kenney. Reston, Va.: The Council, 1991.

North Carolina School of Science and Mathematics. *New Topics in Secondary School Mathematics: Matrices.* Reston, Va.: National Council of Teachers of Mathematics, 1988.

Price, Jack, and J. D. Gawronski, eds. *Changing School Mathematics: A Responsive Process.* Reston, Va.: National Council of Teachers of Mathematics, 1981.

Romberg, Thomas. *Rational Assessment for Meaningful Reform.* Proceedings of the Ohio Mathematics Educators Conference. Columbus, Ohio, 1990.

Stenmark, Jean Kerr, et al. *Assessment Alternatives in Mathematics: An Overview of Assessment Techniques That Promote Learning.* Berkeley, Calif.: EQUALS Project, University of California at Berkeley, 1989.

Swasy, Alecia, and Carol Hymowitz. "The Workplace Revolution." *The Wall Street Journal Reports, Education.* 9 February 1990, pp. 6–8.

Thompson, D'Arcy Wentworth. "On Magnitude." In *The World of Mathematics,* Vol. 2. Edited by James E. Newman, pp. 1001–46. New York: Simon and Schuster, 1956.

Wheatley, Grayson. MAPS: Mathematical Applications through Problem Solving. Unpublished material. West Lafayette, Ind.: Purdue University, 1988.

Wirszup, Izaak, and Robert Streit, eds. *Developments in School Mathematics Education around the World: Applications-oriented Curricula and Technology-supported Learning for All Students.* Reston, Va.: National Council of Teachers of Mathematics, 1987.

AN ANNOTATED BIBLIOGRAPHY OF CORE RESOURCES

The following is an annotated resource list of materials of possible help in developing a core curriculum and establishing a rationale for why such a curriculum is necessary. It is organized by title of the work in the following classifications for easy reference: *Other Addenda Booklets (9–12)*, *Materials Documenting Changing Needs*, *Assessment and Evaluation Materials*, *Core-related Materials from Other Countries*, *Innovative Curriculum Resources*, and *Innovative Software Tools*.

Other Addenda Booklets

Algebra in a Technological World. M. Kathleen Heid et al. Reston, Va.: National Council of Teachers of Mathematics, forthcoming.

This book examines how technological tools such as graphing utilities, symbol manipulators, spreadsheets, computer algebra systems, and multiple-linked representation programs support a new vision of algebra. It addresses recommended changes in content as well as in teaching, learning, and assessment. Examples and activities illustrate the effective use of technology in mathematical modeling and in investigating families of functions, matrices, linear and nonlinear systems, algebraic structures, and dynamical systems.

Connecting Mathematics. Gary W. Froelich et al. Reston, Va.: National Council of Teachers of Mathematics, 1991.

A compendium of examples and classroom-ready investigations that illustrate how new content, such as data analysis and matrices, and new perspectives on familiar content, such as rectangular coordinates, functions, mathematical reasoning, and problem solving, can serve to connect traditional topics that remain important but that students often see as isolated. Throughout, there is a focus on connections among mathematical topics and between mathematics and real-world situations.

Data Analysis and Statistics across the Curriculum. Gail Burrill et al. Reston, Va.: National Council of Teachers of Mathematics, forthcoming.

Examples and activities that illustrate how to integrate statistical concepts into the traditional high school mathematics curriculum form the essence of this booklet. Intended for use with students, the units include exploring data, interpreting data presented in the media, topics from algebra and geometry involving statistics, and an introduction to using models to fit curves. There are suggestions for giving and grading student projects, strategies for assessment, and many practical ideas for use in the classroom.

Geometry from Multiple Perspectives. Arthur F. Coxford et al. Reston, Va.: National Council of Teachers of Mathematics, 1991.

This volume links coordinate and transformation approaches to geometry with the synthetic approach commonly found in today's geometry classrooms. It features investigations and real-world applications that lead to student-generated theorems and that build geometric intuition. Possible contributions of Logo turtle graphics, drawing and measuring computer utilities, and graphing calculators to learning and teaching geometry are addressed.

Materials Documenting Changing Needs

A Challenge of Numbers: People in the Mathematical Sciences. Bernard L. Madison and Therese A. Hart. Committee on the Mathematical Sciences in the Year 2000, National Research Council. Washington, D.C.: National Academy Press, 1990.

The companion document to *Everybody Counts,* this booklet documents the needs in the mathematics pipeline at the college and university levels. Chapters 2, 3, and 4 are particularly appropriate for 9–12 programs in preparing students for the transition to collegiate-level work and opportunities.

Chaos: Making a New Science. James Gleick. New York: Penguin Press, 1988.

A penetrating chronicle of how computers have enabled scientists to use simple recursive techniques and quadratic functions to describe patterns in real-world phenomena that are chaotic. Both the ways in which the scientific world initially resisted these ideas and the implications for mathematics educators to rethink the role of a classical preparation in the calculus for advanced study are worthy of reflection.

Everybody Counts: A Report to the Nation on the Future of Mathematics Education. Mathematical Sciences Education Board. Washington, D.C.: National Academy Press, 1989.

A concise but encompassing assessment of the system for developing mathematical expertise in this country characterizing the changing value and character of mathematics as a discipline, misconceptions about factors affecting achievement, and actions necessary for moving programs into the twenty-first century.

Mathematics: A Project 2061 Panel Report. David Blackwell and Leon Henkin. Washington, D.C.: American Association for the Advancement of Science, 1989. (See *Science for All Americans.*)

Prepared by the mathematics panel for the Project 2061 report, *Science for All Americans,* this booklet is an excellent analysis of mathematics as a discipline, its processes, subject areas, language, and relationship to other disciplines.

Renewing U.S. Mathematics: A Plan for the 1990s. National Research Council. Washington, D.C.: National Academy Press, 1990.

This is the David II Report, which follows the 1984 David I review and recommendations concerning the strength of the mathematical sciences in this country. The needs and resources addressed in this report concern primarily the graduate levels of mathematics education.

Science for All Americans: A Project 2061 Report on Literacy Goals in Science, Mathematics, and Technology. Washington, D.C.: American Association for the Advancement of Science, 1989.

Presents an expanded description of the need for scientific literacy for citizens of the next century, detailing a common core of learning in science, mathematics, and technology for all young people.

The Underachieving Curriculum: Assessing U.S. School Mathematics from an International Perspective. Curtis C. McKnight et al. Champaign, Ill.: Stipes Publishing Co., 1987.

Characterizes poor achievement results for the United States in the Second International Mathematics Study compared to most other countries in the study. It recommends five steps for the renewal of U.S. mathematics, including restructuring of the curriculum and reconsidering differentiated curricula (tracking).

Workforce 2000: Work and Workers for the Twenty-first Century. William B. Johnston and Arnold E. Packers. Indianapolis, Ind.: Hudson Institute, 1986.

Characterizes trends in work force needs into the twenty-first century and describes the worker skills required for the types of jobs available. Changes in demography, work force constituency, types of jobs, pay scales, and educational requirements provide a rationale for a core curriculum to meet the country's needs.

Assessment and Evaluation Materials

Assessment Alternatives in Mathematics: An Overview of Assessment Techniques That Promote Learning. Jean Kerr Stenmark, the EQUALS staff, and the Assessment Committee of the California Mathematics Council *Campaign for Mathematics.* Berkeley, Calif.: EQUALS Project, University of California at Berkeley, 1989.

A highly teacher-usable listing of alternatives in assessing student achievement to promote learning. Sample assessment items, assessment issues, and advantages of alternative techniques are discussed in this forty-page booklet. An ideal staff development tool for expanding colleagues' perspectives.

From GATEKEEPER to GATEWAY: Transforming Testing in America. Chestnut Hill, Mass: National Commission on Testing and Public Policy, 1990.

A general reference on the issues involving the need to transform testing in the U.S. from overreliance on group-administered, paper-and-pencil, multiple-choice instruments to methods more appropriate for developing America's human resources. Materials in this booklet primarily address policy-level considerations.

A Question of Thinking: A First Look at Students' Performance on Open-ended Questions in Mathematics. Sacramento, Calif.: California State Department of Education, 1989.

A report on the use of open-ended questions on the twelfth-grade test of the California Assessment Program in 1987–88. A rationale for including open-ended questions and rubrics for scoring such items are included in addition to large-scale student responses and implications for improving classroom instruction.

The STATE of Mathematics Achievement: NAEP's 1990 Assessment of the Nation and the Trial Assessment of the States. Washington, D.C.: National Center for Education Statistics, 1991.

A comprehensive national summary of achievement data on NAEP testing in mathematics at grades 4, 8, and 12 in 1990. Results are analyzed by level of outcomes, demographic factors, curriculum patterns, instructional approaches, and other classifications. Calculator and computer use, characteristics of mathematics teachers, and other factors affecting instruction are also included. There is a companion document listing data for each state (e.g., *The STATE of Mathematics Achievement in WYOMING*).

Core-related Materials from Other Countries

Better Mathematics. Afzal Ahmed. London: Her Majesty's Stationery Office, 1987.

A discussion of the Low Attainers in Mathematics Project (LAMP), including goals, teacher development, curriculum development, assessment, and parent involvement.

Developments in School Mathematics Education around the World: Applications-oriented Curricula and Technology-supported Learning for All Students. Izaak Wirszup and Robert Streit, eds. Reston, Va.: National Council of Teachers of Mathematics, 1987.

Proceedings of the University of Chicago School Mathematics Project International Conference on Mathematics Education.

Developments in School Mathematics Education around the World: Applications-oriented Curricula and Technology-supported Learning for All Students, Vol. 2. Izaak Wirszup and Robert Streit, eds. Reston, Va.: National Council of Teachers of Mathematics, 1990.

See especially "The Reform of Mathematics Education at the Upper Secondary School Level in Japan" by Fujita Hiroshi, Tatsuro Miwa, and Jerry P. Becker, which describes the reform of the secondary mathematics curriculum in Japan to a core curriculum through grade 10. The program is developed along a Core-and-Option (COM) format.

Mathematics Insight and Meaning. Jan de Lange Jzn et al. Onderzoek Wiskundeconderwijs en Ondersijscompetercentrum (OW&OC), Rijksuniversiteit Utrecht, Holland, 1987. rite Vakgroep OW&OC, Tibberdreef 4, GG Utrecht, the Netherlands.

Includes examples of the new secondary school education in the Netherlands, which took place with the introduction of the new curricula in 1985. It describes content and addresses issues such as the role of pure mathematics over applied mathematics in the curriculum.

Mathematics Program in Japan. Research Center for Science Education, National Institute for Educational Research, 6-5-22, Shimomeguro-ku, Tokyo 153, Japan, 1989.

A description of the new Japanese mathematics curriculum (kindergarten through upper secondary school levels) being phased in from 1990 to 1994.

New Mathematics Curriculum. Japanese Ministry of Education, Tokyo Books, 1984. English translation by UCSMP, 1990.

A complete translation of seven Japanese schoolbooks for grades 7–11 done under the supervision of Izaak Wirszup of the resources component of the University of Chicago School Mathematics Project. Available for $10.00 each or $60.00 for the series. Inquiries should be addressed to Izaak Wirszup, Director, Resource Development Component, Department of Mathematics, University of Chicago, 5734 University Avenue, Chicago, IL 60637. Note that the Japanese curriculum is currently being rewritten. [See also *Mathematics Program in Japan.*]

School Mathematics in the 1990s. Geoffrey Howson and Bryan Wilson. International Committee on Mathematical Instruction (ICMI) Study Series. Cambridge: Cambridge University Press, 1986.

A discussion of mathematics in a technological society, mathematics content, teachers and classrooms, research, and other issues.

Shadow and Depth. (Translated into English by Ruth Rainero.) OW&OC, Rijksuniversiteit Utrecht, Holland, 1987. rite Vakgroep OW&OC, Tibberdreef 4, GG Utrecht, the Netherlands.

One of a series of units written for the Dutch secondary school mathematics curriculum. The units are reality-based and involve in-depth problem solving at a challenging level. Most are in Dutch. *Matrices* by Jan de Lange and Martin Kindt is also translated into English.

Socio-Cultural Bases for Mathematics Education. Ubiratan D'Ambrosio. n.p.: UNICAMP, 1985.

A look at ethnomathematics issues including mathematics literacy, new roles for teachers, out-of-school experience, and curriculum development and research priorities.

Statistical Investigations in the Secondary School. Alan Graham. Cambridge: Cambridge University Press, 1987.

A prestatistics course that offers no formulas. Rather, it presents simple graphical methods and a commonsense approach in an activity format to introduce students to what statistical thinking is all about. Includes a computer software pack.

Innovative Curriculum Resources

The Art of Problem Posing. Stephen I. Brown and Marion I. Walter. Hillsdale, N.J.: Lawrence Erlbaum Associates, 1990.

Helps teachers and students take known problems and modify them to create new problems. An excellent resource for enrichment topics.

Changing School Mathematics: A Responsive Process. Jack Price and J. D. Gawronski, eds. Reston, Va.: National Council of Teachers of Mathematics, 1981.

Contains many helpful articles about the directions in which mathematics programs need to go as well as considerations involved in the process of change.

Discrete Mathematics across the Curriculum, K–12. 1991 Yearbook of the National Council of Teachers of Mathematics. Margaret J. Kenney, ed. Reston, Va.: The Council, 1991.

For those looking for a comprehensive definition of discrete mathematics, this is the ideal resource. Plentiful examples and ideas for integrating discrete mathematics topics at the 9–12 (and other) grade levels are included.

Mathematics across the Curriculum and *Introduction to Algebra and Statistics.* Philip C. Green et al. Dayton, Ohio: EFA and Associates, 1988.

This two-course set of materials is the product of the Ohio MATH Project to develop nonremedial materials for non-college-intending students that are representative of the mathematics they will need. The first course emphasizes a problem-solving approach to applications of previously studied mathematics ideas through a project focus that will span 3–5 days of learning time. The second course, suitable for implementing core ideas, is built primarily around statistics applications to graphing real-world data, which eventually lead into algebraic ideas as ways to describe the relationships conveyed by graphing data.

On the Shoulders of Giants: New Approaches to Numeracy. Lynn A. Steen, ed. Washington, D.C.: National Academy Press, 1990.

Five authors present their views of themes around which to reorganize the structure of the mathematics curriculum: *dimension, quantity, uncertainty, shape,* and *change.* Authors show how these fundamental ideas are deeply rooted in the development of mathematical ideas across many age levels and levels of abstraction. The purpose of the book is to stimulate thinking about innovative curriculum approaches rather than to offer a prescription for how to achieve reform ideas.

Reshaping SCHOOL Mathematics: A Philosophy and Framework for Curriculum. Mathematical Sciences Education Board. Washington, D.C.: National Academy Press, 1990.

A framework for reforming mathematics curricula in this country, the document addresses principally two issues from *Everybody Counts* and the NCTM *Curriculum and Evaluation Standards:* (1) changing perspectives on the need for mathematics, the nature of mathematics, and the learning of mathematics; and (2) changing roles of calculators and computers in the practice of mathematics.

The Secondary School Mathematics Curriculum. 1985 Yearbook of the National Council of Teachers of Mathematics. Christian R. Hirsch, ed. Reston, Va.: The Council, 1985.

Two sections containing several chapters each are particularly apt: "New Curricular Directions" and "Innovative Three- and Four-Year Programs." New directions, including integrated programs and strengthened programs for the not-necessarily-college-intending students, are addressed.

Taking Charge of Change. Shirley M. Hord et al. Alexandria, Va.: Association for Supervision and Curriculum Development, 1987.

This book for educational change facilitators teaches the Concerns-based Adoption Model (CBAM), which emphasizes change as a process, not as an event, and presents the personal and contextual nature of change.

*Teaching and Learning Mathematics in the 1990s.*1990 Yearbook of the National Council of Teachers of Mathematics. Thomas J. Cooney, ed. Reston, Va.: The Council,1990.

Ideas for expanding mathematical instruction to new populations through more effective teaching practices and the use of technology are key sections of this yearbook. Assessment issues and cultural and contextual factors in instruction are additional sections worthy of consideration.

Wisconsin Applied Mathematics Program. Madison, Wisc.: Department of Public Instruction, 1990.

A curriculum guide for the schools of Wisconsin to present a core-based program to non-college-intending students.

Innovative Software Tools

MathCAD 2.0. Student ed. MathSoft, Inc., One Kendall Square, Cambridge, MA 02139.

The student edition of MathCAD is an inexpensive, shortened version of the extremely powerful, professional numeric mathematics package. It is based on a notebook or scratch pad metaphor, allowing text and mathematics to be combined efficiently into a single 2-page document. The software evaluates algebraic, trigonometric, and statistical expressions and generates tabular and graphical displays. Approximate numerical solution techniques are used to solve equations. The student edition offers matrix operations limited to either five rows or five columns and systems of equations with at most ten unknowns, but it contains the full complement of tabular, graphical, and approximate numerical operations from summations, derivatives and integrals, and regression statistics for data analysis.

Computer Laboratories

MultiPurpose Lab Interface Program. Vernier Software, 2920 S. W. 89th Street, Portland, OR 97225.

The MultiPurpose Lab Interface Program is available for both Apple and IBM computers. The easy-to-install hardware and menu-driven software allow students to use their computer to gather and analyze data. The probes gather and display the data from experiments with temperature, pH, and force. The software offers an oscilloscope mode that allows students to experiment with sound. The data collected can be imported into MathCAD for analysis, if desired.

Personal Science Laboratory. International Business Machines Corporation, 1133 Westchester Avenue, White Plains, N.Y.

The Personal Science Laboratory is a combination of software and easily connected hardware that allows students to conduct experiments and collect data with the computer as the investigative tool. Sensors and probes to measure temperature, light, distance, and pH connect tho the computer and offer immediate graphical and tabular feedback as the experiment is in progress. The menu-driven software allows students from elementary school through high school to gather, plot, and analyze data. The collected data can be imported, with additional software, into the Mathematics Exploration Toolkit, if desired.

Geometry

Geometric PreSupposer. Sunburst Communications, Inc., 101 Castleton Street, Pleasantville, NY 10570-3498.

The Geometric PreSupposer allows students to build their understanding of geometric shape and form by offering quick and easy constructions and measurements of the basic geometric figures, segments, triangles, polygons, and circles. Concepts such as parallelism, similarity and congruence, and perimeter and area can be explored and developed.

Geometric Supposer Series. Sunburst Communications, Inc., 101 Castleton Street, Pleasantville, NY 10570-3498.

This series, Geometric Supposer: Triangles, Quadrilaterals, and Circles, allows students to formulate and test their own conjectures. The software allows the students to draw and take measurements quickly and accurately on the geometric figures under study. Midpoints, angle bisectors, and perpendicular and parallel line segments are all easily constructed. Data may be collected and compared as the figures are altered and the constructions are repeated automatically. These programs are available for Apple, Macintosh, and IBM computers.

Symbol Manipulation

Derive. Soft Warehouse, Inc., 3615 Harding Avenue, Suite 505, Honolulu, HI 96816-3735.

Derive is a menu-driven, symbolic algebra program, which offers 2- and 3-dimensional graphing capabilities. Derive generates exact analytic solutions to equations or approximate numeric solutions. A split-screen option allows for simultaneously viewing 2- and 3-dimensional graphs as well as algebraic expression and equations. The software supports matrix algebra with matrices whose dimensions are limited by the space available to store the elements. The software also allows for iterative processes and recursive functions.

Mathematics Exploration Toolkit. International Business Machines Corporation, 1133 Westchester Avenue, White Plains, NY.

The Mathematics Exploration Toolkit is a command line, symbolic algebra program. The software is presented in forty-column mode for ease of viewing in classroom demonstrations. Mathematical expressions are presented in standard form and are easy to read. MET offers quick and easy 2-dimensional graphics. The zoom facility is particularly useful. Tables of values are quickly generated so that the student can see simultaneously algebraic, graphical, and numeric interpretations of functions and equations.

Theorist. Prescience Corporation, Conduit Mathematics Software, The University of Iowa, Oakdale Campus, Iowa City, Iowa 52242.

Theorist uses the efficiency of the Macintosh point-and-click technology to ease the entry of equations and expressions. It allows the user to simplify expressions and solve equations symbolically. Theorist offers graphs of parametric equations, graphs in polar coordinates, and 3-dimensional surfaces. It also offers symbolic matrix algebra. Each work session is stored in a notebook to be saved and recalled, if desired.

Matrix Algebra

Matrices. National Council of Teachers of Mathematics, 1906 Association Drive, Reston, VA, 22091.

The software, which comes with the NCTM booklet *New Topics for Secondary School Mathematics: Matrices,* is a numerical matrix algebra package. The software is designed to allow for ease of entry, editing, and evaluating sums, products, powers, and inverses of up to 15×15 matrices. Row reduction and full pivots are also performed. The software offers forty-column mode for classroom demonstrations with a large monitor or projection screen.

See also Theorist, Derive, and MathCAD for matrix operations.

Statistics

Data Insights. Sunburst Communications, Inc., 101 Castleton Street, Pleasantville, NY 10570-3498.

Available for Apple IIe, IBM, and Tandy computers, the software enables students to work easily with large sets of data. With little instruction, they can display data in a variety of plots and calculate statistics. Every screen can be printed and used for analysis outside the classroom. Using this as a tool, students have time to experiment with data and can concentrate on the analysis necessary to solve problems. A manual of student materials is part of the package.

Data Models. Sunburst Communications, Inc., 101 Castleton Street, Pleasantville, NY 10570-3498.

Available for the Macintosh, this easy-to-use software enables students to enter paired data and fit lines and curves. It also provides equations, correlation coefficients, residuals, and their graphs. Students can investigate data transformations in both tabular and graphic form.

Statistics Workshop. Sunburst Communications, Inc., 101 Castleton Street, Pleasantville, NY 10570-3498.

Available for the Macintosh, the software allows students to explore histograms, means, box plots, and scatterplots and, a unique feature, to investigate categorical data. It provides "stretchy" histograms to allow students to manipulate distribution and explore what happens to the descriptive statistics. Students are able to move fitted lines visually through the graphs of paired data to determine relationships.